Algerian Memories:

a bicycle tour over the Atlas to the Sahara

by

Fanny Bullock Workman

Rise of Douai

Copyright © 2016 Rise of Douai

ISBN-13: 978-1539691242

ISBN-10: 1539691241

About the Authors

Fanny Bullock Workman (January 8, 1859 – January 22, 1925) was an American geographer, cartographer, explorer, travel writer, and mountaineer, notably in the

Himalayas. She was one of the first female professional mountaineers; she not only explored but also wrote about her adventures. She set several women's altitude records, published eight travel books with her husband, and championed women's rights and women's suffrage.

Born to a wealthy family, Workman was educated in the finest schools available to women and traveled in Europe. Her marriage to William Hunter Workman cemented these advantages, and, after being introduced to climbing in New Hampshire, Fanny Workman traveled the world with him. They were able to capitalize on their wealth and connections to voyage around Europe, North Africa, and Asia. The couple had two children, but Fanny Workman was not a motherly type; they left their children in schools and with nurses, and Workman saw herself as a New Woman who could equal any man. The Workmans began their travels with bicycle tours of Switzerland, France, Italy, Spain, Algeria and India. They cycled thousands of miles, sleeping wherever they could find shelter. They wrote books about each trip and Fanny frequently commented on the state of the lives of women that she saw. Their early bicycle tour narratives were better received than their mountaineering books.

At the end of their cycling trip through India, the couple escaped to the Western Himalaya and the Karakoram for the summer months, where they were introduced to high-altitude climbing. They returned to this then-unexplored region eight times over the next 14 years. Despite not having modern climbing equipment, the Workmans explored several glaciers and reached the summit of several mountains, eventually reaching 23,000 feet (7,000 m) on Pinnacle Peak, a women's altitude record at the time. They organized multiyear expeditions but struggled to remain on good terms with the local labor force. Coming from a position of American privilege and wealth, they failed to understand the position of the native workers and had difficulty finding and negotiating for reliable porters.

After their trips to the Himalaya, the Workmans gave lectures about their travels. They were invited to learned societies; Fanny Workman became the first American woman to lecture at the Sorbonne and the second to speak at the Royal Geographical Society. She received many medals of honor from European climbing and geographical societies and was recognized as one of the foremost climbers of her day. She demonstrated that a woman could climb in high altitudes just as well as a man and helped break down the gender barrier in mountaineering.

In June 16, 1882 married William Hunter Workman, a man 12 years her senior. He was also from a wealthy and educated family, having attended Yale and having received his medical training at Harvard. In 1884 they had a daughter, Rachel.

William introduced Fanny to climbing after their marriage, and together they spent many summers in the White Mountains in New Hampshire; here she summited Mount Washington (6,293 feet or 1,918 metres) several times. Climbing in the Northeastern United States allowed Fanny to develop her abilities together with other women. Unlike European clubs, American climbing clubs in the White Mountains allowed women to be members and encouraged women to climb. They promoted a new vision of the American woman, one who was both domestic and athletic, and Workman took to this image with enthusiasm. By 1886, women sometimes outnumbered men on hiking expeditions in New England. In her paper on the gender dynamics of climbing in the region, Jenny Ernie-Steighner states that this formative experience shaped Workman's commitment to women's rights, pointing out that "no other well-known international mountaineers of the time, male or female, spoke as openly and fervently about women's rights". However, both of the Workmans disliked the provincial nature of life in Worcester, where they resided, and yearned to live in Europe. After both Fanny's and William's fathers died, leaving them enormous estates, the couple embarked on their first major European trip, a tour of Scandinavia and Germany.

INTRODUCTION

ALGERIA with its three provinces – Oran, Alger and Constantine – is about twelve hundred and twenty-five kilometres long, and stretches southward from the Mediterranean into the Sahara some six hundred and sixty kilometres. It consists of a series of table-lands flanked by the elevations of the Atlas mountains, which, under various names, extend through the entire country, sending spurs in every direction. The coast appears from the sea as a bold wall, rising frequently into peaks of quite respectable height.

Both mountains and plateaus increase in altitude as the interior is penetrated, and from west to east. The table-lands vary in height from two hundred to eleven hundred metres. The highest mountain in the province of Oran, the Touabet of the Tlemcen range, attains an elevation of sixteen hundred and twenty-one metres; that in the province of Alger, the Lalla Kredidja of the Grande Kabylie range, two thousand three hundred and eight metres; and the Chelia of the Aures range, in the province of Constantine, reaches two thousand three hundred and twenty-eight metres. The mountains in general are not remarkable for boldness of outline, but in certain regions, as in the Djurjura and the Chabet-el-Akhra, the weird shapes of the towering peaks, the sombre desolation of the verdureless craggy masses, the tremendous gorges with rushing mountain torrents, combine to form scenery of the first order.

The climate of Algeria varies with the altitude, as elsewhere, but as a rule it is mild in winter and spring. Although perhaps uncomfortably warm in summer, it cannot be called tropical, except south of the Atlas mountains in the Sahara. On the coast the temperature averages in January, February and March about 50° Fahr. The nights and mornings are cold, so that fires are desirable, though frost is seldom seen; but at mid-day, in the sun, one can be entirely comfortable. From April on, the heat increases, and by the end of May the traveller usually considers the time fitting to move farther north.

The temperature during the summer months reaches 80° and 90° Fahr., and remains quite constantly at these figures. On the table-lands the cold is greater in winter, and the heat in summer, than on the coast. The climate here can be called agreeable only a couple of months in the year. Snow and severe cold are often experienced. In the last week of March we saw ice an eighth of an inch thick one morning in the Province of Constantine, and found the climate rough and disagreeable.

South of the mountains in the Sahara, as at Biskra and Laghouat, the climate is rather milder in winter than on the coast, and the number of sunny days much greater. Biskra, which can now be reached by rail, is being advertised as a winter health resort. Still even here the wind, although not so cold as on the high lands, blows

4

often with such force, and raises so much dust, as to make open-air life uncomfortable for invalids.

The higher mountains are covered with snow until the last of May, and even later. The rainy season usually lasts from October to the end of February, and sometimes till the end of April. From October to May cold north-west to north-east winds prevail. After May the Sirocco is not infrequent, raising the temperature as high as 113° Fahr., and bringing from the Sahara dust and other characteristic phenomena.

The vegetation is similar to that of Sicily and the southern parts of Europe. Among the fruits are the lemon, orange, almond, mandarin, fig, olive, nectarine, peach, plum, apricot, grape, date, pear and apple. The forest trees correspond to those of other parts of the South Temperate Zone. Especially noticeable are the cork oaks, the giant ashes of the Kabylie, the cedars and olives.

The best time to make a trip through Algeria is from about the middle of March to the end of May. Before the first-named time the sky and roads are too uncertain to admit of successful travel. After the end of May the heat becomes too great for comfort. The vegetation is not far enough advanced before the middle of April to give the best impression of the country. The end of May is about as early as any climbing among the higher mountains can be indulged in on account of the snow, which, not being permanent, is too dangerous to venture upon.

Railways have been built connecting Oran on the west and Constantine on the east with Algiers, together with several shorter side branches running to strategic points. By the use of these the traveller may see a great deal. But many of the most interesting and original districts, and the finest scenery, are to be reached only by carriage, on horseback, or the bicycle. He who would know the land as it is, must be prepared to travel a good deal in one of these three ways. We chose the last, which mode of travel we have employed in Europe for several years and have found most independent and satisfactory.

The French appreciate the value of good roads, and since their occupation of Algeria they have devoted considerable attention to the construction of proper means of communication between important points. Thanks to their efforts, Algeria possesses a system of highways that will compare favourably with those of European countries, excepting England, France, Italy and Norway. Some stretches, especially in the mountains, are equal to any to be found in any country, being in every respect first-class. Such are the roads from Miliana over the mountains towards Algiers, that through the Gorge of Chabet-el-Akhra, supported almost entirely on a bed of solid rock, and that south from Khroubs to Batna and El Kantara.

A large number of roads built of limestone are hard and excellent when dry, but soften and become slippery when wet. The roads from Oran towards Tlemcen are of this character. Others, having a superficial covering of ordinary clay soil, are fair when dry, but when wet become utterly unridable with the bicycle, on account of the tenacious, glue-like quality of the mud, which adheres to and clogs the wheels. In the desert the roads degenerate into camel and mule tracks, in places quite ridable, and in places rough. After rains the large feet of the camels, sinking into the soil a foot or more, leave the surface in an impassable condition.

The grades are generally easy, like those in France, so that many of the ascending stretches can be ridden, and all of the descending, few of the latter requiring the use of the break. The French do not carry their roads directly over a steep hill as do the Germans, but wind it so as to keep the grade easy.

The method used in the construction of roads is to spread a thick layer of broken stone over the whole road bed and roll it in. A difficulty is experienced in the character of the soil, which, when soaked by rain, allows the stones to sink below the surface, so that a road when first made deteriorates more or less rapidly, which renders a re-treatment with stone necessary. So far as observed, the roads, which had received a second layer of stone, remained smooth and hard.

Algerian memories;
a bicycle tour over the Atlas to the Sahara

CONTENTS

CHAPTER I

ORAN AND THE MOURDJADJO

IT was the second day out from Marseilles. The Ville de Bone had made a good inroad on the forty hours' run to Oran. The first dinner had not proved a success. The passengers who had appeared at table disappeared about the time of the third course. On this second afternoon all were more cheerful and able to gaze with considerable interest at the picturesque outlines of the Balearic Isles, which, as passed about sunset, were resplendent in purple light.

At dinner, our little ships' company began to get acquainted. It consisted of the captain, first officer, doctor and four passengers, the wife of the English consul at Grenada, a lieutenant of the English army on his way from India to Gibraltar, and ourselves. We two Americans, bound on a cycling trip to Algeria, excited a mild sort of interest in the captain, a man possessed of the sang froid of a Frenchman combined with the indifference of the colonist in Algeria. The French resident of that country appears to have thrown off all interest in his native land as well as in the land of his adoption, and, for that matter, in life itself.

After answering his questions, we, in turn, questioned him as to what Oran could offer for sights.

'Rien,' he replied, 'rien du tout;' and with the natural shrug of the shoulders. 'It is simply a French city.'

'But,' we urged, 'out of seventy-four thousand inhabitants there are only nineteen thousand French. There must surely be something of Oriental character and custom among the remaining inhabitants to attract the visitor?'

His only reply was the eternal shrug, which can mean so much, but which usually means so little in a Frenchman.

Finding no knowledge was to be gained as to the city and people, we tried again in a different direction.

'There is a fine view, is there not, captain, from a mountain called the Mourdjadjo, which everyone ought to climb?'

He did not look like a man fond of scenery, and his answer strengthened that impression.

'Je n'en sais rien de cela. But there is a Fort Santa Cruz that all strangers visit. I live in Oran, but have never been there myself.'

'Oh, we know about that. It is all in the Guide Joanne,' we replied. 'But what we wish you to tell us about is the Mourdjadjo, and the famous view upon Oran.'

'I do not know it; it does not exist.'

'Not exist!' we repeated with surprise.

We had chosen Oran as the starting-point for our journey, chiefly because of its proximity to Tlemcen, but the prospect of a visit to the Mourdjadjo furnished an additional inducement.

A German writer of ability has described the view from this mountain upon Oran as being more impressive than the noted one from the Weissthor upon Macugnaga, eight thousand feet below, and this had lingered hauntingly in our minds. To be told now by a resident of Oran there was no Mourdjadjo was indeed a disillusion.

We asked if he knew Tlemcen. Yes, it was interesting, because situated in a garden of verdure.

'But you had better go by rail.'

'Why, are the roads bad?'

'No, excellent, but the Arab dogs are dangerous, and attack strangers in troups.'

This was not encouraging, particularly to a woman whose skirts had previously sustained manifold injuries at the mouths of Swiss and Italian dogs. Even the thought of our revolvers and dog-whips did not quite dispel the gloomy impression produced by this statement of the phlegmatic skipper.

'How shall we find the people?' we next asked.

'Not bad as far as Tlemcen. Beyond that I do not know.' He warned us, however, against the Maltese, and said, – 'If you wish to know anything, ask the Arab rather than the Maltese.'

As it proved, it was the Arab and no the Maltese who was on hand during our wanderings.

On February 14th, about noon, the Ville de Bone entered the harbour of Oran under the bluest of African skies, and in a short time we were on the way to the Hotel Continental, followed by two Arabs, in heik and burnous, pushing the machines.

Oran is the least interesting of, and has less of the characteristics of Eastern life than any of the Algerian cities. Out of a population of seventy-four thousand, only about nine thousand are Arabs, the rest being composed of French, Spanish, Gipsies, Jews and other nationalities, mostly of the lower classes. The architecture is as polyglot as the inhabitants. Although, in the squares and market-places, Oriental types may still be seen, the commonplace European is the prevailing figure met with not dressed in the latest style, and one realises how permanently here the Arab has stepped aside for the European.

The artistic sense is offended by the half European appearance of many of the natives, who have discarded their national costume and adopted the sack coat and long trousers, which are not becoming to their athletic figures. The effect is rendered still worse by the retention of the fez.

The oddest of all costumes is that worn by the negro labourers and box-carriers, the lowest element of the population all over Algeria. This consists of an imitation, in common burlap or sacking, of the large cloaks worn by the better classes. In place of the pointed hood, which is often wanting, the upper edge is thrown over the head and fastened with a cord around the neck. As this cloak wears out it is patched with pieces of all kinds and colours sewed on with string, one layer superimposed upon another, until all traces of the original texture have disappeared, and a garment is produced that would put to shame the many-coloured coat of Scriptural renown. The dark face of the wearer, with his glittering eyes, white teeth and good-natured grin, and his bare nether extremities projecting below the ragged border of this covering, form a picture that, once seen, can never be forgotten.

Having finished our survey of the city without finding even an interesting mosque, we concluded to start for the Mourdjadjo. Guide Joanne, published by Hachette, the only reasonably reliable guide of Algeria, states that Fort Santa Cruz stands on the summit of the Mourdjadjo, and then proceeds to describe two excursions, one to Santa Cruz, the other to the last-named mountain, which is mentioned as rising behind Santa Cruz, and affording a better view of the country.

Like lucid explanations in regard to other places were destined at times to make our path a thorny one, and we often longed for a concise, direct Baedeker. No one can express himself more lucidly, and no one can compose more indirect and labyrinthine sentences than a Frenchman. Unfortunately, the man who compiled Joanne belongs to the latter class.

The people whom we met seemed imbued with the guide-book idea that Santa Cruz stood on the Mourdjadjo, and accordingly directed us toward the fort. As the dark bald head of the mountain could be seen looming up behind the fort, we knew they were mistaken. The difficulty was to find the right path. Finally, after a half-hour of fruitless wandering, a woman gave us the proper direction, and in an hour and a-half we were at the top, from which, the day being hazy, we did not get the distant view, and saw nothing of Almeria and the coast of Spain mentioned in the guide.

From the edge of the plateau which formed the top, we looked down upon Oran. While the view of the city, eighteen hundred feet below, bordered by the deep blue sea is very effective, only the enthusiasm of a German imagination could make it so telling on the mind as that upon Macugnaga, eight thousand feet below the Weissthor.

A small koubba stands upon the summit occupied by an old marabout. When we asked permission to enter his sanctum, he pointed to our feet. Relieving them of boots, we stepped for the first time into a Mohammedan sanctuary, which differed from the many mosques afterwards visited, in that it was absolutely clean and a good bit smaller than a state room on an ocean racer – so small, in fact, that the ancient priest slept on a straw mat in a niche in the massive brick wall. A taper was burning, and a collection of embroidered silks hung about, such as one sees in Algerian mosques. Beyond these there was nothing except a little hot-water heater, the presence of which suggested that possibly the old man might sometimes solace himself with a cup of tea or a glass of grog.

We gave him some sous which, notwithstanding his religious ardour, he was not loath to accept.

Before descending the mountain we had a foretaste of what was in store for us in the matter of African lights, in the wonderful purples and sapphire blues that settled over the distant mountains.

CHAPTER II

TLEMCEN, CITY OF MOSQUES

AN accident, which required several days to repair, having happened on the passage to one of our wheels, we decided to act on the captain's advice and take the rail to Tlemcen, beginning the bicycle journey a few days later from Oran.

Tlemcen originally consisted of two towns, Agadir and Tagrart, which were afterwards united and surrounded by one wall. The former has wholly disappeared. The latter, the site of the present day Tlemcen, was founded in 1080 A.D. by the Almoravides. Different dynasties followed up to the time of the Almohades, under whom the city began to develop. After the latter came the Abd-el-Ouadites, who were constantly at war with the Merinides, who finally annexed Tlemcen to their empire, and although they held it only twenty-two years, it reached a higher state of prosperity and wealth under them than under any previous dynasty.

The Abd-el-Ouadites again obtained the upper hand, and extended their empire over the territory now comprising the Provinces of Oran and Alger. Under them Tlemcen attained the zenith of its glory; the number of inhabitants reached one hundred and twenty-five thousand, it became the home of poets, and a seat of learning and scholarship. It was called the city of mosques, counting sixty-one within its walls, many of which, built by sultans and deys, were examples of the richest Moorish architecture.

Its decline began early in the sixteenth century, and was partly brought about by the conquest of Oran by the Spaniards. After this it was occupied alternately by Turks and Moroccans until 1836, when Marshal Clausel directed his expedition against it. It was permanently occupied by the French in 1842.

At the present time Tlemcen has twenty-nine thousand five hundred inhabitants, of which thirty-six hundred are French. It has been sacked and destroyed so many times that little is left of its ancient glory. Its palaces and monuments have disappeared, its sixty-one mosques have dwindled to three, only the shell remains. Still, that remnant is very Eastern in character, and the French have made but little impression either on the architecture or the people.

The Grande Mosqué, as is often the case with the grandes mosqués of Algeria, is hardly worth unbuttoning one's shoes for. The view from the tower is good. The smaller mosque of Djama l'Hassen has an interesting interior, with handsome arches falling upon onyx columns. The walls are finely sculptured, and the ceiling of cedar is a relic of Moorish carving, upon which may be seen traces of the original polychromatic colouring.

The mosque of El Haloui has a picturesque courtyard with fountain, and inside eight superb marble columns, the capitals of which rival any at the Alhambra as specimens of Arab ornamentation. Except in these and a few other instances, searching for art in Algerian mosques, shuffling over dusty carpets in sandals worn by everybody, is dull work, which is relieved of its monotony only by the spectacle of Musselmen, in the performance of their devotions, bowing their foreheads to the floor in front of the mihrab.

In a place like Tlemcen, where everything is so totally different from what is to be seen in Europe or America, the street life possesses a peculiar attraction, and the visitor may profitably stop here for some days to observe it. The weekly markets bring in large numbers of people from the country, who display articles for sale in the squares and market-place. Fruits, oil, hay, live stock, cereals, shoes, clothing and other necessities of life are bought and sold. Caravans also come in from Morocco under charge of native Moors and desert negroes. The opportunity here offered the kodakist is a rare one. Unfortunately for him, however, the various objects of his ambition dislike being photographed, and make off with themselves as soon as he levels his instrument. Much patience and some finesse are required to obtain good results.

The modern Mechouar, or, as the French call it, Place d'Armes, arranged as a spacious shady promenade, contains little to recall the splendid palaces where sultans ruled with an iron hand, and the marble courts where poets sang to the soft music of fountains in the palmy days of the Abd-el-Ouadites. Around it stand to-day the prison, the hospital, the barracks and the magazines. It is still the centre where the race in power places its government buildings, but the modern utilitarian buildings do not suggest the graceful structures that once occupied this spot.

Yet, towards sunset on a clear, spring-like February afternoon, as the setting rays fall aslant the faded mosque of Djama-el-Mechouar, where Almohades, Merinides and Turk, each in his century, came to pray, we feel again the spirit of the Mechouar, and seem to see noble Moors on their way to prostrate themselves at the mihrab of the mosque. And in fact, as the shadows lengthen, the old square becomes alive with a very pageant of interesting figures come for their evening airing.

There are Moors such as are seldom seen in Algiers, for Tlemcen is only seventeen kilometres from the border of Morocco, and Arabs lean and tall, with peaked faces, and negroes broad and brawny, with massive Ethiopian features, and lastly the Jew, who adds just the brilliant colouring needed to set off the scene in his red, blue or yellow burnous and bright turban, sharply contrasting with the white drapery and simple heik, adorned with twisted yarn coils, of the Arab.

It is like the effect of fine music on the mind to watch them as they walk forward and back, or stand talking in groups, or, better still, sit about on the benches. Who ever looks twice at men, women or children sitting in the Boston Common? At Tlemcen, every man thus seated is a subject to be framed and sent to the Paris Salon. Some of the older Jews, engaged in earnest conversation, gesticulating with uplifted arms and flashing eyes, would make Shylocks such as modern stages cannot produce. Others, sitting in silence, in various attitudes, closely muffled in their warm, soft cloaks, the folds of which fall carelessly around them, present figures no less interesting.

When the dusk deepens and they gradually move away, leaving the square deserted, the dreary French prison fades with the light, and in its place, under

13

the bright African night, rises a pale, Moorish outline of delicately-twisted columns supporting sweep after sweep of crenellated archways, into which disappear the fascinating subjects of our day-dream in the Mechouar, and with them departs the spirit of the Merinides, the Abd-el-Ouadites, and the Moors.

Unique, and unlike any other in Algeria, is the Jewish quarter, still occupied by the Jews in much the same condition in which it was in former times. They allow visitors to examine their houses freely. These houses have two stories below the level of the street, and one or more above. The rooms are small, and built into the walls as alcoves from the passage-ways, and have no furniture, except a low frame, over which a covering is thrown, which serves as chairs and bed. They are grouped around an inner court similar to, but smaller than, the Moorish patio, which supplies them with light and air.

Queer little spiral stone stairways, hardly wide enough for a single person, connect the different floors. The underground portions, which were well inhabited, were damp and dark, and appeared about as comfortable as subterranean prison cells. These houses were built during the reign of the sultans and deys, who cruelly persecuted the Jews, and it is said to be owing to their malice that the latter were obliged to construct and live in such dwellings. On the outside of many of the houses, near the entrance, were impressions of an open hand in blue and red, made by dipping the hand into paint and pressing against the wall. This is regarded as a sign of good luck.

Cooking is done over small portable fire-pots, which are carried from room to room. Everywhere a strong odour of burning wood was noticeable. The day of our visit, the women were busy scrubbing and whitewashing the courts and alcoves in preparation for the Sabbath, but they were good-natured, and willing to show us their houses, and appeared highly pleased when we gave the children a few sous.

JEWISH AND ARAB CHILDREN, TLEMCEN.

The Jewish children are especially attractive, dressed as nowhere else, in red, yellow and purple-figured calicoes, which they drape over their shoulders with charming effect. They also wear jaunty little pointed caps. We made several attempts to kodak them, but in vain, for as soon as they discovered our intention they would scamper away. Finally, an offer of money induced some to yield to our wishes, though some of the most attractive would not, probably from fear of our tiny Kodak, which they doubtless regarded as some strange and evil thing.

The Arab quarter is extensive, and teems with Arabs, often of splendid form – pictures of grace and strength in their flowing burnous. The women, or Mauresques, cannot be said to impress one with the idea of grace, as they move about in baggy trousers, their heelless shoes clattering noisily on the pavements. In Oran and Tlemcen the women veil the entire face, except about the half of one eye, which half they evidently use to the best advantage.

Interiors in the Arab quarter were also visited. There only women are allowed to enter. The houses seen showed a greater degree of comfort than those in the Jewish quarter. Instead of being greeted at the door by gaudily-dressed girls with laughing, uncovered faces, as among the Israelites, the visitor passes through a long, narrow, silent corridor, which leads to a cheerful patio, with prettily-decorated tiled walls and the customary fountain in the centre. On one side an arch forms the entrance to the living-rooms, where women and children are seen lying about on rugs, or sitting cross-legged, some eating, some sewing, some lazily rocking cradles. Here, where their faces were uncovered, no beauties were observed.

They were friendly and pleasant, but perfectly inactive. Perhaps four rooms on a patio are thus occupied. The numerous children crowd around the visitor, begging for money, in this respect differing from the Jewish children, who, to our surprise, although willing to take money, did not beg.

Next comes the bath, which is interesting as showing that, however luxurious the bathroom of a high-born Eastern lady may be, the bath room of the average family is a very every-day, uncomfortable place. On the occasion of our visit, young girls, superintended by an old woman, were performing their ablutions in a common room, the bath-tubs used being wooden, and of the most ordinary character.

A prominent feature of the street life is the serpent charmer from Morocco, who, with a bag filled with repulsive, venomous-looking reptiles as his stock-in-trade, furnishes as much entertainment to the populace as the Polichinelle to the frequenters of the Champs Elyseés. When the performance is about to begin, a trio of attendants beat a dismal refrain on a large tambourine and a couple of tamtams, which is repeated at intervals until the close.

We join the circle which is formed around the serpent-charmer, in this case an old man with white hair and beard. He takes two large snakes from the bag, holds them up and talks to them, then laying one on the ground, beside a boy who acts as assistant, he places the other inside the bosom of his shirt. He raises his arm, and as the sleeve falls away, the serpent is seen to emerge and coil itself around it. It draws its head back threateningly. The spectators hold their breath. An Arab zouave next us whispers, 'Regarde, tu vera, il va mordre.'1 Suddenly it seems to strike the forearm. A blood-red fluid flows from the spot and trickles over the hand. A murmur of horror passes around the circle. The old man now disengages the snake from his arm, lays it on the ground and walks around, gesticulating, pointing to the wound and talking vociferously. The lookers-on assent with frequent exclamations to what he says, and although they must know, having seen like exhibitions all their lives, that the charmed serpent does no injury to the performer, they appear as earnestly interested as if they expected to see the latter writhing the next moment on the ground in agonies of death.

At this point the boy passes around the tambourine, which receives a goodly number of sous. The operator then opens the bag, shows that it contains several more serpents, and causes the boy to thrust one leg up to the hip into it. The boy's face assumes an expression as if he does not exactly fancy the neighbourhood into which his leg has strayed. As he sits on the ground the other serpent crawls up and winds itself about his neck. We pass on.

The country around Tlemcen merits investigation, affording some charming excursions. One sunny afternoon we found ourselves en route for the Bois de Boulogne, a grove of magnificent centenary olives. It begins about a mile from the town, and extends up the side of the mountain Lella-Setti, upon which the city is built. Through this olive grove rough, stony paths lead in different directions, and at every turn, in the early spring, descend rollicking streams, small tributaries of the Oued Kala. An English traveller, who had never been in Algeria, once said to us that anyone wishing to see olive trees should go to Spain and Corfu. When cycling on the coast of Calabria we descended one day from the mountains to Palmi, through a fine olive plantation, which had since lived in our memory as surpassing, in the size and beauty of its trees, any elsewhere seen; but we now had to confess that the olives of Palmi paled before those of Tlemcen. Had the English traveller been in Algeria, it is possible his opinion might have been modified.

Scattered about among the trees are a number of koubbas, both ancient and modern. The latter are distinguishable by their whitewashed exteriors. The former, which are half in ruins, are much more picturesque. One in particular, with indented arches, attracts the attention, erected to the memory of the savant marabout El Tiyar, who died at Tlemcen in 1295. He was the pride of his age, and numerous miracles are attributed to him, among which was the power of appearing, by enchantment, first in one place and then in another. His name signifies the flying man, and had he lived in Nibelungen days, he would have been said to possess the Tarnhelm. All the old koubbas cover the remains of honoured Tlemcen marabouts, or were dedicated to their memory.

Travellers only care for them in so far as they serve to impress the influence of other days upon the mind. A ruined koubba or mosque has the advantage over other ruins, that when in its doubtful beauty one is striving to recall the spirit of past ages, one has but to look around to see, crouched against the temple wall, in white robes, Mohammed himself, gazing up with lustrous, inquiring eyes.

A RUINED KOUBBA, TLEMCEN.

As, on February 17th, we rest under the olives, on the grass sprinkled with violets, primroses and anemones, listening to the babbling brook at our feet, and watching the passing Arabs, some on foot, some on horseback, with their veiled women sitting behind, it is like a scene from the 'Arabian Nights,' and an effort is required to recall the fact that, one week before, we were living under the leaden winter skies of Germany.

Spring was not merely in the air. It was spring during the few days spent at Tlemcen. We had reason to remember it, for although from that time on, except on the higher plateaus, cherries and apricots bloomed about us, and the vegetation was as far advanced as in May in Germany, the air was often filled with the breath of winter, and, particularly among the mountains, uncomfortably cold.

Beyond the Bois a road leads to the village of El Eubbad, perched on the side of the mountain, and surrounded by gardens that rise in terraces and are watered and kept green by innumerable rills. Olive, fig and pomegranate trees

17

covered with ivy overhang this small paradise. Here is the mosque of Sidi Bou Moudin, formerly one of the finest in Algeria and still containing rich columns, mosaics and carvings. The minaret was once entirely covered with faience work, some of which remains. In its high situation it must have been a conspicuous landmark, as seen from below in the valley of the Tafna.

Connected with the mosque is the 'Medersa,' or college for higher studies. When we visited it the exercises seemed to consist chiefly of the sing-song repetition of the Koran usually heard in Arab schools.

Between El Eubbad and Tlemcen are the ancient and modern cemeteries. The former is completely neglected; the latter contains some fine cypresses.

Another excursion is to the site of the former city of Mansoura, three kilometres west of Tlemcen, a commanding situation with extensive view. In 1300 the city was large and flourishing, and, like Tlemcen, filled with great palaces and gardens. It was noted for its strong fortifications and active trade. In 1306 the Black Sultan took up his abode there and began the siege of Tlemcen, which lasted two years.

Five centuries have passed since the destruction of the city, and nothing is now left except the immense quadrangle formed by the ruined towers and remains of the outer wall, the foundations of a mosque and one face of a minaret, which still retains traces of the magnificence which must have reigned in Mansoura. To-day, ploughed fields and waving grain cover most of the area then occupied by human habitations. Olive and fruit trees grow in profusion around the old towers, softening and beautifying their time-worn outlines.

With regret we left Tlemcen and the comfortable one-storied Hotel de France, where we were well taken care of at moderate prices.

CHAPTER III

ORAN TO AFFREVILLE OVER THE OUARSENES AND THE CHELIF — AFRICAN DOGS.

WHILE at Oran, among other preparations for the tour we tried to find suitable maps of the country. Bicycle maps, so far as we could learn, did not exist, neither were good road maps to be had. We had to content ourselves with a very ordinary, not too accurate, general map published in Paris, which answered the purpose fairly well, though defective in details. When, on various occasions, we attempted to supplement its deficiencies by inquiries as to the existence, character and condition of roads, we found that information on such topics was difficult to obtain in Algeria, and, when obtained, was in very few cases reliable. Even a prominent bicyclist in Algiers, when consulted in regard to the road from Batna to Biskra, gave a description which proved to be wide of the facts, as afterwards experienced.

The noon of February 21st saw us strapping on the last packages. Salem, the handsome Arab porter, took the final orders about forwarding our trunk, and accompanied by the good wishes of the proprietor and a group of natives and Europeans, who appeared interested in our undertaking, we wheeled the rovers out of the courtyard of the 'Continental,' and mounted for our journey of over fifteen hundred miles through Algeria.

It seemed a wonderful ride, that first afternoon run of eighty-one kilometres from Oran to Perregaux. That was probably because it was the first ride in Africa, for, when analysed, it was, except for being rather more novel, very like a half-day's spin in France or Italy. The country was rolling, the air mild, trees and flowers in bloom, the roads fine, and offering the advantage of not being cut up by vehicles. After leaving Oran we met only Arabs and negroes, on foot, or riding on donkeys and horses. We delighted them and they charmed us, and it was with mutual satisfaction that the American and Arab met en route that day.

They were good-natured and orderly, and, when we several times rode through large companies of them, although they laughed and accosted us, they did so civilly, and were neither rude nor coarse. Even when their animals, frightened at the unusual sight, shied up a bank or into a field, they took it in good humour. Once an Arab was thrown from his horse, but he did not seem disturbed by the mishap. We were struck with the difference in temperament between the Arab and the Sicilian, as we recalled the various occasions on which, when in Sicily, we had been the unintentional cause of unhorsing the latter, who, although not apparently injured by the fall, would usually curse us vociferously with fierce gesticulations, as we rode on.

It has been our experience that horses, oxen and mules are much more liable to be frightened by a woman on a bicycle than by a man. Dogs also bark at the former more frequently. It may be that dogs, which seem to regard themselves as a sort of special police, consider women out of place on a wheel, and in need of correction.

In Sicily and Southern Italy, on dismounting in a town, we were immediately surrounded by a motley and noisy rabble, which accompanied us until we left it. On similar occasions, in Algeria, a few Arabs would gather slowly about at a respectful distance, evidently interested, but entirely silent and undemonstrative, never offering to touch the machines. The Arab, as a rule, is lazy and not fond of work or overexertion, yet he sometimes displays both activity and endurance. On that ride to Perregaux a young man ran along with us for two kilometres or more, as we rode at a fairly rapid pace.

We reached Perregaux, a town of fifty-eight hundred inhabitants, about six o'clock, and found a very comfortable inn, where a passable dinner of six courses was served for two francs each. A table-d'hôte of some pretension was generally attainable, even in small places, where the other accommodations were of the simplest character, at about the price mentioned. Occasionally a charge of three and a-half francs would appear upon the bill, which did not necessarily imply that the dinner was any better, but might rather be taken as an indication that the attention of the hostess had been called to the practice prevalent in Europe, of making a higher charge to English travellers than to those of other nationalities. What the charge would have been had she known we were Americans, we would not venture to guess. Suffice it to say, it is always to the advantage of his purse, if an American abroad can pass for an Englishman, which, on the ordinary routes of travel, it is next to impossible for him to do.

As is the case in the south of France, the inns of Algeria are frequently kept or managed by women, and the chambers are cared for by men, who make excellent chambermaids. While yet uninitiated in the customs of the country, we sought one evening the bureau of an inn, and asked an intelligent-looking woman sitting there, where the host was. 'C'est moi,' was the reply, and she proved fully equal to the demands of her position.

Soon after leaving Oran, opportunities of verifying the truth of the captain's statement about dogs began to present themselves, and long before our Algerian tour was finished, we were thoroughly convinced that the facts, in this instance at least, had been correctly stated. As we passed farmhouses and native habitations, the dogs would rush out at us, sometimes singly, sometimes in twos and threes, barking furiously, snapping and showing their teeth in a most threatening manner. These dogs are shaggy, gaunt, wolfish-looking beasts, with long, sharp noses and glaring eyes, are taught to be suspicious of strangers, and are rendered more savage by being half-starved. The most ferocious are kept chained or shut up during the day, but it is never safe to approach a house, unless armed with a stout cane. What would have happened to us had we not been provided with steel-cored whips, it is not difficult to predict. To say the least, we should speedily have become candidates for the Pasteur treatment. To increase the efficiency of these, we had taken the precaution to fasten good-sized shot on the snappers. This worked well on the dogs, but was detrimental to the whips, as the weight of the shot under constant use caused the snappers to break off. The idea then occurred to us to provide the lower end of the whips with six wire barbs similar to those used on barbed fence wire, each projecting three-eighths of an inch. One blow with the whip thus armed was usually sufficient. The barking would change instantly into a short, sharp yelp, and the dog would slink off conquered. The sudden transition from an attitude of confident attack to one of ignominious defeat was most amusing.

Later on, between Algiers and Constantine, as we were passing an Arab village a little off the road, one evening after dark, we were startled by a tremendous barking. Of a sudden, at least fifty dogs broke out in full chorus and barked as if they would tear everything around them to pieces. Whether we were the cause of the deafening din we did not know, nor did we know whether they were chained, but the prospect of being attacked in the dark by these howling fiends was not reassuring. They did not molest us, and we once more breathed freely as the sounds grew fainter behind us.

The further journey to Affreville was uneventful. Between the towns only Arabs were seen. On market days we met them in large numbers, going into or out of the town where the market was held. Most of them were on foot, but the richer class rode on horses or mules.

In the Ouarsenes mountains we passed over long reaches of sparsely-inhabited country, where not even the usual shepherd boy was visible. Coming around the barren mountain side, into view of the plains of the Chelif, a remarkable landscape lay spread before us. On one side, the mountains overhanging in jagged outlines the valley stood out clear but black as midnight, shadowed by heavy clouds, while beyond, the wide plain swept for miles, an oasis of spring green illuminated by softest sunlight.

The African lights were a constant delight, whether studied on cloudless or overcast days. Except near the coast, fogs were rare, as were low tones in colour, clear outlines, sharp contrasts and strong colours prevailing.

CHAPTER IV
MILIANA AND THE TROGLODYTES OF THE ZAKKAR

FROM Affreville in the plain, one may go either to Teniet-el-Had among the mountains, to the South, to visit the giant cedars, or climb to Miliana, a town seven hundred and fifty metres above the sea, and, like Olevano and Taormima, renowned for its situation. We did not consider it prudent to go to Teniet-el-Had, as it lay high; the season was early and we had but little warm clothing.

Miliana is nine kilometres from Affreville and four hundred and fifty metres above it, the rise being continuous. A negro youth was engaged to push one of the rovers. We were somewhat anxious as to the result, as he had probably never beheld such an instrument of locomotion before, which, in the hands of one unaccustomed to its use, is liable to be unmanageable. More than once we had seen on similar occasions, boy and bicycle, becoming hopelessly entangled, roll together on the ground. But our anxiety was groundless. He neither dropped the rover nor knocked his feet against the pedals, and appeared well satisfied at the end of the climb, which was accomplished in about two hours, with the reward he received.

Miliana stands on the side of a grand mountain called the Zakkar. The view from the terrace or promenade, off upon the Plains of the Chelif and the distant Teniet-el-Had mountains, reminds one of that from Olevano towards the Sabine Hills. When we were there, although a week still remained to the end of February, the weather was bright and the air mild. Trees and flowers were bursting into bloom, and the grass was green upon the whole slope, watered by rivulets born of the fast-disappearing snow-fields of the Zakkar. In this we were favoured, as there is never any certainty, at this time of the year, of meeting mild weather at such an altitude, even in North Africa. Miliana is better visited in April.

The cities and large towns of the interior of Algeria, such as Tlemcen, Mascara, Blida, Setif, Constantine, Batna and others, are fortified by walls of brick or stone built immediately around them, with gates where the roads enter. These walls serve to accentuate the impression produced on the mind by the enclosed buildings, and the visitor may easily imagine himself transported backward a few centuries in time. They have been built during the last forty years, and their character was determined by the necessities arising from the French occupation. They are of about the height and strength of the twelfth-century walls of Europe. A glance shows they would not, for an hour, withstand ordinary field artillery, and that they are intended as a protection against infantry only.

When the French subdued the country the population was disarmed, particularly of artillery. Probably, not a piece of artillery of any kind could be found to-day in Arab hands. In case of insurrection, therefore, any garrison, having to contend against small arms only, and these of an inferior kind, would find the walls a sufficient protection. On the other hand, if the Arabs should seize a town, they could not hold it, as the French artillery could speedily destroy both walls and town without any danger to themselves.

Miliana is surrounded by walls reconstructed by the French on the remains of those built by the Romans and Turks. It is an important military point, and has a garrison of four or five thousand soldiers. The Arab troops impress one favourably, being tall, handsome, well-built, neat in dress and modest in behaviour. The infantry wear the Zouave uniform. The cavalry at Miliana, in red trousers and turbans, white frocks and dark blue sashes, mounted on white horses, were very striking.

Although the French do not trust the natives, yet, as a matter of policy, they encourage the enlistment of the latter in the infantry and cavalry service, but not in the artillery. Military service is popular among the Arabs, who enlist voluntarily for a term of four years. They are better fed and cared for, and not obliged to work as much as they would have to do to support themselves. They also enlist to escape from the control of their parents, who can demand their service after they become of age. Re-enlistment is permitted, and many avail themselves of the opportunity to serve several terms. We saw soldiers at least forty years old serving with the colours.

The Arab soldier can rise only to the rank of lieutenant. If he wishes to advance beyond this grade, he must be baptised and become a French citizen. Military service is beneficial to the natives, in that it teaches them habits of order and obedience to authority, and makes them familiar with the French language and a civilised mode of life, all of which is also to the advantage of France, as they thereby become better subjects and more reconciled to French dominion. They form an organised body, which is available in case of civil accident. In the forest fires, which recently caused much damage in Algeria, the soldiers rendered efficient service in subduing the flames.

Although so early in the season, we decided to attempt the ascent of the Zakkar, being already half-way up at Miliana. Two reasons influenced us in this decision – first, the one which always obtains with nature enthusiasts – the prospect of a comprehensive view; and second, the desire to see some of the wild dwellers on the mountain, who live in caves and openings in the rocks. The Zakkar, fifteen hundred and eighty metres in height, is a bold, craggy mountain with two peaks. We took a bright young Arab to show us the way, as, beyond a certain point, the path ceased. To that point it was rough enough, and we marvelled at the ease and rapidity with which he climbed in light sandals where we felt the stones through stout boots. An active scramble of an hour brought us to the tiny house of an old marabout, standing on a projecting point, from which an enchanting view of Miliana, Chelif and a rolling background of mountains was unfolded. We had expected to gain information from the marabout in regard to the amount of snow higher up, but unfortunately he was absent. The following day being fête des femmes, all the Arab women of Miliana were to ascend the mountain as far as this point, and he was away making arrangements for their reception.

We went on, and after a while came to patches of snow, which gradually increased in size and depth until at length, when all three stood knee deep in wet snow, the uselessness of further attempt became apparent, and we realised that African mountains are no freer from snow, at a certain height, than those of some other countries.

After descending a little, the guide took us to another part of the mountain inhabited by the cave-dwellers. The Zakkar is here covered with jutting crags, forming little caves in which whole families live. In these, in the absence of chimneys, the atmosphere was stifling from the smoke of burning roots and twigs. In these worse than desolate dwellings were no household utensils, except a pan or two, in which the galette, or hard bread made of flour and water, which forms the chief article of diet, was prepared. The people support themselves by carrying down wood and charcoal to the weekly market at Miliana.

They are timid, shy, and so afraid of Europeans that when, in hunting, the French officers come on them, they beg for mercy and offer to surrender the little they possess to avoid molestation. Their clothes are the veriest tatters.

Material for covering the women's faces is a luxury not to be thought of, and these are left exposed to the public gaze.

Towards sunset we returned to Miliana. Not much was left of the guide's sandals, but he did not seem concerned, and drank absinthe to our health at the Café de la Terrace, while we drank sirop de grenadine to his. It would be well for Algeria if temperance societies could wage war against the sale of the former beverage. The Arabs have become much addicted to the absinthe habit, as well as the French colonists, and its pernicious effect is very evident. We were told, by good authority, that many colonists, who are at first thrifty, in a few years become shiftless and good for nothing.

We found syrup of grenadine, or pomegranate with water, a delicious and refreshing drink. It was to be had in almost every café. Water taken with this syrup never seemed to affect us disagreeably, as it sometimes did when taken alone. The Hotel d'Isly at Miliana is one of the best hotels of the Algerian interior.

CHAPTER V
THROUGH THE VALLEY OF THE METIDJA – BLIDA, GORGE DE CHIFFA, ALGIERS

WE first appreciated the extent of the Zakkar when, in the early morning, we rode along its flank at least ten miles before beginning to descend. Then, leaving it behind, we went down into a deep valley, and again ascended over a long range of hills, from the top of which the sea came into view for the first time since leaving Oran. Again, a long down over an excellent road, and we came whirling into the blooming valley of the Metidja, whence we bore down upon Blida, home of the mandarin. From the snow of the Zakkar we had come, in one day, into the summer-land of oranges, lemons and figs, and now only fifty kilometres of valley, and a climb over the plateau of the Sahel, separated us from Algiers. But Blida and the Ruisseau des Singes were to keep us a day or two yet from the climate-seeker's paradise, and, quite content with the glorious run from Miliana, we turned into one of the very comfortable Blida hotels. Scarcely had quarters been selected, when a plate of mandarins was brought in for the tired bicyclists, with the compliments of the proprietor. We were not accustomed to such attentions from Algerian hosts, who generally contented themselves with questions as to the cost of our rovers; hence it was all the more appreciated in this case.

The following day we rode to the Gorge de Chiffa and the Ruisseau des Singes. The Gorge, although pretty and picturesque, is much over-rated, and in itself not particularly worth a visit. If, however, the monkeys are really seen clambering among their native rocks, one feels repaid for the trouble of coming. The sight is rare of late years, and sometimes for ten days not a monkey appears. Having heard the early morning was the best time to see them, we arrived soon after seven o'clock at the inn which guards the entrance to the

Ruisseau. Here, in conformity with a notice stating that, if visitors did not take some refreshment, an entrance-fee would be charged, we cheered the heart of the innkeeper by refreshing ourselves.

Entering the rift through which the brook ran, we walked along the overhanging path for some time, with eyes fixed on the opposite side. At length we espied five large monkeys springing from rock to rock, then others smaller – twenty to thirty in all. They had been down to the brook to drink, and were returning up the mountain-side. They stopped frequently to pick berries, and as they ate them, sitting in giddy places, they looked across at us in the most comical manner. They had no tails. It is said to be from this race of monkeys that the organ-grinders of Europe are largely supplied.

Blida is an uninteresting half French, half Arab town, important to the French, agriculturally, being located in the most fertile part of the Metidja. It has large orange, mandarin and lemon groves, and exports some five million oranges annually to Paris. The Bois Sacré of ancient olive trees is one of the most beautiful in Algeria. It is a Mecca for the Musselmen of that region, who are daily seen prostrating themselves at the ornate koubbas, in reverence to the memory of the learned marabouts, whose spirits still seem to hover among the silvery branches of the great spreading trees.

On the route to Algiers we passed through Boufarik, celebrated for its large native cattle-market on Mondays, when about five thousand Arabs from the Metidja assemble there. It is difficult to imagine that Boufarik, with its wide, well-cared-for streets and allées of superb plantains, was, until the time of the French occupation, regarded as a dangerous pest hole. The Arabs in those days held a weekly market there, but broke up as early as possible, in order to escape the malarial emanations at evening.

We wheeled on to Algiers, over the Sahel plateau, famous in guide-book lore, but to us, after Miliana and the Zakkar slopes, rather tame. On arrival in Algiers we were besieged by boot-blacks and street gamins whenever we stopped to inquire the way, and also before the hotel. They popped out from behind every pillar, becoming more importunate, as their number increased, with their 'Cirer, m'sieu, cirer.' While we were unstrapping our luggage, numerous boxes of yellow paste were thrust in our faces with the cry, 'Cirage jaune, moi j'ai le meilleur,' and when the hotel porter drove them off, they scattered, only to return to the attack before we had entered the door.

An excellent opportunity is afforded in Algiers to study the demoralising effect almost invariably exerted on a semi-barbarous people by contact with a so-called civilised one. The latter usually conducts itself in such a manner that its vices are imitated sooner than its virtues by the weaker race over which it has acquired control. Added to this, in Algiers, is the customary pernicious influence exercised by travellers. In following the ordinary routes of travel, one sees practically nothing of the real characteristics of a people. A spoiled wine gives no measure of the quality of the same when good. He who judges the South Italians by the lower classes around Naples, errs greatly in his judgment, and he likewise who judges the Algerians by those he meets in Algiers. The natives are naturally temperate, quiet, retiring, modest and polite. In Algiers, the street population has learned to be forward, bold and impertinent, and has become as great a nuisance as that of any travelling centre.

Ten years ago, there was something to write about in Algiers; to-day there is less, and ten years hence there will be nothing. The city is rapidly becoming Europeanised. A large part has been remodelled and rebuilt in conformity to modern requirements, and the process is still in progress. The inhabitants have largely adopted European dress and manners.

The situation of the city is fine, but not so beautiful as that of Bougie, seldom visited by the tourist. After the mosques and palaces have been examined, but little remains to detain one. It is not to be wondered at that visitors flee to the Mustapha Superior for relief from the noise of the narrow, badly-lighted streets of the inner city. The Kasbah still offers some interesting points. One street in particular is worth seeing, the only one that is in the same condition that it was before the coup d'evantail, where the walls of the houses above the first storey project, being supported on short, slanting poles, and finally, at the top, almost meet their opposite neighbours. Old quaintly-decorated doors are here and there to be found, which one studies with a sense of regret at the prospect that another year or two may see them torn down by the ruthless hand of the French builder, for the Arab house, like the Arab costume, is bound to disappear from Algiers.

The mauresque is best seen in the few streets of the Kasbah. She wears the same costume as the mauresque of Oran and Tlemcen, except that her forehead and eyes remain uncovered. The latter, dark, and often pretty, seem to gaze wistfully after the European woman who visits her quarters, with a look in which envy and hopelessness are combined. Through letters of introduction, we were invited to the Moorish villa of a French official and his wife living in the suburbs of Algiers. This was constructed in the usual manner, with one storey, the rooms opening on a patio, and with heavy walls two feet or more in thickness. Over and around it trellised vines grew luxuriantly. The salons were lighted by recessed windows, and ornamented with arches set off by Oriental rugs and Parisian furniture. The owners said they could occupy the villa throughout the summer without feeling the heat unduly, a fact which explains why the French in Algeria often build their new houses with thick walls, in the Moorish style.

CHAPTER VI
PALESTRO – THE LONELY COLONISTS OF BENI MANÇOUR – BORDJ BOU ARERIDJ

ABOUT four hundred and seventy kilometres separate Algiers from Constantine. The first hundred kilometres the route passes through an undulating, highly-cultivated country, commanding grand views of the snow peaks of the Djurjura. Menerville, Palestro and Bouira are all well situated on the border of North Africa's grandest scenery, which a month later we were to visit. The inns proved very comfortable, and most cosy was the scene in the dining-room at evening. While rather obliged to partake of that nuisance, the table-d'hôte, serving à la carte not being very well understood in the small towns, each party has its own table. In the salle à manger gathered the English or French traveller, the officers, who always dine at the best inns, the commis voyageurs, and not least well represented, if last, the dog, and sometimes the cat. By the time the fish was served, eight or ten of the former would be on hand, mostly large dogs belonging both to the host and guests of the house. They went from table to table, sitting up with a comical gravity, and watching with eager eyes every morsel put into one's mouth. Yet they were quiet and patient, did not whine or beg, and were adept at catching any tit-bit that might be thrown them.

We were favoured with three days of uninterrupted sunshine after leaving Algiers, the temperature, on starting at six o'clock in the morning, was 50° or 52° F., and by eleven the air had the warmth of a European June day. The atmosphere was marvellously clear, and objects could be seen distinctly at great distances. There was no hazing up towards mid-day as is so often the case in Sicily and South Italy. The mountains and plains were wrapped in a deep blue or purple light, which never varied, but was just as entrancingly beautiful at high noon as in the early morning. After the lapse of months, memory still recalls vividly the intense blues and purples on those and on succeeding days, bathing landscapes, sometimes barren and desert, with a glory which, without this divine lighting, would not have been theirs.

Before reaching Palestro, we rode through the Gorge of Palestro on a perfectly-constructed road, running high over the rushing river between limestone cliffs, whose bases were covered with prickly pear and other subtropical vegetation. The whole place is so much more striking than Chiffa, that it is surprising the guide-book should devote so much space to the latter, and so little to Palestro.

We arrived one noon at the little station of Beni Mançour. As we were placing our rovers against the wall outside, preparatory to going in to see what the larder would offer, a pale, sallow Frenchwoman came to the door and advised us to take in all our luggage, or it might be stolen by the Kabyles, pointing as she spoke toward half a dozen shaven-headed natives coming up the road. We followed her advice, and while eating the eggs, and contemplating the raw artichokes she served, listened to her dismal account of their neighbours. She, her husband and children lived there alone, surrounded by Kabyles. Although they barred the windows and bolted the door, as is the custom with the French colonists whenever they left the house, nearly all their household gods had been stolen, from lamps and mattresses to the Frenchman's new fusil, over the loss of which he sadly shook his head. 'Ils ne sont pas mauvais, les Kabyles, mais ils sont voleurs,' he said.

Between the thieving Kabyles, and the fever to which they were victims throughout the summer, this couple appeared to have little reason to be pleased with the land of their adoption. The woman looked pityingly at us as we prepared to leave, saying, –

'Are you going up there?'

'Yes,' we answered cheerfully, as we glanced towards the splendid road winding upward into the barren hills. 'It is up hill and lonely, but we do not mind that.'

'C'est plus que desert c'est triste, et les gens sont mauvais,' she answered.

That the people were bad we had no proof, although we were warned not to be out after nightfall, but that the country was triste we had abundant opportunity to realise, and that afternoon, before dark, the full dreariness of the region settled like a weight upon our spirits.

The road led up and down among the most desolate clay hills, washed in all directions by the winter rains; some entirely devoid of vegetation, others covered with silver-barked pines. Then it ran between mountains, composed of crumbling argillaceous rocks and slate of imperfect formation, with strata, perpendicular or inclined at an acute angle, the harder parts standing above the rest in points and ridges. It was a land of cheerless distances, more desert-like in some respects than the real Sahara. Now and then we saw perched on a bleak hill-top a small Kabyle village. People we rarely met, passing only here and there a single shepherd or stray Kabyle standing motionless in the fields, staring at us with wondering eyes. Even the camel, met with at intervals since leaving Algiers, failed here.

After a long ascent, as the shadows were lengthening, we came to the hamlet of El Achir, and found nine kilometres still remained before we could find shelter for the night, but the road was good and the distance was covered in half an hour. With real pleasure, as darkness was setting in, we hailed the scattered lights of the town of Bordj bou Areridj – a name not particularly suggestive of home comforts. We were, however, well taken care of.

CHAPTER VII
THE AFRICAN FARM – STALLED IN THE MUD – SETIF TO CONSTANTINE

THE weather changed during the night, and the sky looked threatening when we started in the morning, but we determined to push on, hoping to reach Setif before rain should set in. The character of the country was rather less dreary than on the previous day. After riding four hours it began to rain hard. Fortunately, we espied, a short distance ahead, a group of buildings, surrounded by a high rectangular stone wall, with one entrance toward the road, the only human habitation within several miles. Outside the wall was a cluster of Arab tents. Here we sought shelter.

As we passed through the entrance we were greeted by three fierce Arab dogs, which barked furiously and showed their white teeth, but were prudent enough to keep out of reach of our whips. A fourth dog, confined in a small shed near by, threw himself savagely against the door in the attempt to get at us. We were afterwards informed that this beast was so dangerous that he was always shut up during the day. The buildings consisted of barns and sheds projecting from the wall, which formed one side of each, and one rather more pretentious than the rest appeared to be the abode of man. The noise brought to the door of this last a French colonist, his wife and two children. They received us courteously and invited us in out of the rain. The room into which we entered served as kitchen, dining and general living room, and was generously shared with the chickens and dogs.

The walls and floor were of stone, and the furniture consisted of a table, bench, and two wooden chairs. In place of the kitchen stove, one corner was occupied by a stone fireplace or hearth, over which was the chimney-flue, with a large projecting hood, under which hung some hams in process of being cured. A forlorn place to live in, with no suggestion of comfort. A fire of roots was burning on the hearth, over which the woman was preparing the noon meal, which she cordially asked us to share.

She said they lived there alone, with no companions except the Arabs encamped without the wall, who were in their employ. They did not fear personal violence from the Arabs, but they did not trust them, and always made the outer entrance secure at night, letting loose the dogs in the enclosure as a protection against theft. They called their place 'la ferme,' and said it was the most comfortable one of three they had occupied in Algeria.

With us the idea of a farm is associated with hay and grain fields and orchards, not with a large tract of barren upland covered with scanty vegetation, which is only capable of supporting sheep. Sheep-raising appeared to be the chief occupation on this farm. The Arab shepherds, who looked after the sheep, are paid at the rate of twenty francs per month for every two hundred sheep. This is considered sufficient to supply their few and simple wants, even in the cold Province of Constantine. This provides them with clothing, a breakfast of galette, a dinner of couscous, and leaves something over for absinthe and tobacco. A package of Arab cigarettes costs ten centimes. Only the most capable Arabs are able to take care of two hundred sheep.

After two hours the rain ceased, and, declining the invitation to eat, we prepared to move on. The road was wet, but did not appear unridable, so we bade our hosts farewell and mounted. Alas! we had not proceeded two turns of the wheels before they became clogged and refused to move. Examination showed the tires to be covered to three times their natural size with thick, brown, pasty mud. This we scraped off as well as possible, and tried riding again, but with like result. We were completely stalled, and could not advance a yard. This was our first, but not our last, experience with African mud, slippery as grease, sticky as glue, and drying as hard as stone. The outlook was dubious.

It was well on towards noon, and the distance to Setif was about thirty kilometres – so far as our ability to travel was concerned, it might as well have been a hundred. There was no hope of the road drying that day, and it looked again like rain. The prospect of spending the night in the kitchen of the farm, with the bench and wooden chairs for a couch, was not alluring. The only alternative was to consult the farmer.

He came out, and with the remark, 'Now you will breakfast with us,' shouldered one of the machines and carried it to the house. We attempted to follow suit with the other, which, with the luggage and adherent mud, weighed nearly a hundred pounds, but were glad to relinquish it to an Arab whom the farmer ordered to take it. While the farmer's wife served us with an excellent chicken fricassée, we held a council of war. The only chance of getting to Setif appeared to be to send an Arab on horseback to a village eight kilometres away, where the farmer had an acquaintance who owned a horse and cart, which, if at home, he would willingly put at our disposal.

The Arab was soon speeding over the plain to bring relief. Meantime, out of the seemingly scanty ferme cuisine came, besides chicken, a delicious omelette, bread and butter, wine, café noir and cheese. In an hour and a half the messenger returned and reported the man and cart as en route. He was a long time coming, and we were beginning to be anxious, when the barking of the dogs and the sound of wheels in the court called us out. The cart was small, and some difficulty was experienced in securing our horses properly on it. When this was accomplished, we took leave of our kind friends of the ferme, who, although urged, would accept no compensation for their lavish hospitality. Among our pleasantest recollections are the words – 'Non, non, pas d'argent, ce que nous faisons c'est fait de bon cœur.'

How dull and cheerless it was driving along the road we should, under other circumstances, have ridden comfortably in a quarter of the time. To raise our spirits, the driver regaled us frequently with stories of the racing qualities of his other horse, with which his wife had gone to Setif that day, but these in no way consoled us for being obliged to jog along behind the spavined, worn-out beast he had brought with him. At his house in the village he insisted on stopping for half an hour to feed the horse, during which time he took occasion to drink several glasses to our health. Among twenty or thirty Arabs who gathered around us here, we saw the handsomest man met with in Algeria. They said he was rich, owned much land, and had several wives. He was gentle in manner, had a beautiful smile, noble brow and highly-intelligent face, with penetrating dark eyes that looked as if they had extracted from life the wisdom of the past and future.

He allowed us to kodak him in his clean, soft, fine robes. The day was dark, and the result, on our return to Germany, was found to be a good photograph of the robes, and a dark blot in place of the handsome face. After four hours we reached Setif, thankful, even with delay, to get there, and thus ended our first experience with African mud.

From this point, northward to the sea, southward to the desert, throughout the Province of Constantine, and eastward to Tunis, are everywhere to be found traces of the Roman occupation, many of them in an excellent state of preservation. Following the destruction of Carthage, and the overthrow of the Carthaginian power, 149 B.C., the Romans took possession of the conquered territory and colonised it. They brought their civilisation with them, built roads, bridges and acqueducts, drained marshes and lakes, and introduced their own methods of agriculture, making the fertile upland to yield so plentifully that vast quantities of grain were annually exported to Rome. In fact, these provinces, as well as Sicily, became, and for several centuries continued to be, the granary of Rome. The Romans made the most impression on the eastern parts of Algeria, which most easily submitted to their domination, and adopted their manners and customs, and here the remains are most abundant. The fierce, war-like people of the West, or Maures, were never fully subjugated, and continued a predatory warfare, to which, in the latter days of the empire, the Roman power finally succumbed.

No one can study the Roman remains in different lands without being struck by the thoroughness with which they stamped their own civilisation on all conquered peoples. They did not, as has been so often the case in history, adopt the life and habits of the conquered, but insisted on the conformity of the latter to Roman standards. These were everywhere essentially the same, as may be seen by the similarity of the remains found in different lands. Wherever the Roman went he fashioned his temple, forum, arena, dwelling-house and domestic implements on the same models. Among the Algerian tribes certain Roman customs and household utensils still continue in use. The amphora, with its side handles and pointed end, is seen to-day, especially among the Kabyles, just as it was made in Italy two thousand years ago.

At Setif there is one of the out-of-door museums common in this part of Algeria, containing about one hundred and fifty Roman statues and columns found in the vicinity. They are placed about in a small park without the walls, regardless of the effects of the winter temperature upon them, though Setif lies one thousand and ninety-six metres above sea-level. We were at Setif twice, the first time in March and again in April. The weather was warmer in March, as there was more sunshine than in April. As soon as it rains in Algeria in the spring, the cold becomes disagreeable.

The distance from Setif to Constantine is a hundred and twenty-six kilometres. We left Setif at six o'clock in the morning, and arrived at Constantine at five in the afternoon. The first hundred kilometres were accomplished in seven hours and fifteen minutes, including stops. The last twenty were up hill, and the wind blew in our faces. An hour was taken for lunch and rest. The road was in rather poor condition on account of recent rains. There was nothing especially noteworthy in the nature except the ever-changing panorama of beautiful lights. In place of Arabs in the meadows there were multitudes of storks, standing quite as proudly, though generally on one leg. The minaret of every mosque was surmounted by a nest, and they were seen flying in numbers, a sign, it was said, of an early summer. The stork, when he really sets out to use his wings, can mount to a surprising height and fly rapidly. He does not make as graceful an appearance in the air as when watching for frogs in the marshes.

CHAPTER VIII
CONSTANTINE – RAMADAN CUSTOMS

CONSTANTINE is beautifully situated at an altitude of six hundred metres above the sea, with a commanding view of the surrounding country, which is very fruitful, and especially noted for its oranges, which many consider finer than those of Blida. Its situation has been compared to that of Ronda. It is a natural fortress, and its capture caused the French much trouble and great loss in time and men. One half of its circumference on the east is bounded by a deep ravine, the rocky walls of which fall in the perpendicular from two to three hundred metres. Through this ravine flows the river Roumel, tumbling in cascades over its narrow bed. On the west side the hill slopes off abruptly to the plain below.

Constantine has a history extending back to the time of the Numidians, and has always been a place of considerable importance, having been besieged and taken, according to tradition, some eighty times. Here Sallust, when governor of the province, under Cæsar, had large estates. It has a population of about fifty thousand, of whom twelve thousand are French, and twenty-five thousand Arab. The Arab quarter offers much that is interesting and distinctive. The narrow streets, bordered on both sides by low houses, with small open workshops and boutiques, the occupants of which are busily engaged in the fabrication and sale of all kinds of articles used by the people, also by meat, vegetable and fruit stands, and thronged with purchasers, present an animated scene.

The tanning of hides and manufacture of leather articles forms an important industry, and the number of bootmakers, and other workers in leather, is surprising. There are said to be over five hundred of the former, who provide for the pedal requirements of the whole province. Large numbers are also employed in the manufacture of the woollen and silk fabrics worn by the Arabs. Upon the streets are to be seen the handsome Jewess, with uncovered face and black silk head-covering, the Mauresque, with a blue veil in place of the white one worn at Algiers, the Arab and Kabyle, the Spahis in red burnous, and the Turco in blue tunic and sash.

JEWISH CHILD OF CONSTANTINE.

The most attractive edifice is the palace of Hadj Ahmed, built by the last bey. While not old, it is a fine specimen of Moorish art, having four courtyards filled with shrubs and flowers, surrounded by arcades, with arches resting on twisted marble columns and graceful open-work balustrades. The sides opposite the balustrades have a high dado of yellow and black tiles. Red and blue colouring is also used in the decoration. Standing at one end, and looking through the arcades, the combination of foliage, sunlight, architectural tracery and colour forms a scene of fairy beauty. The palace is better kept up, and makes a pleasanter impression, than the Bardo at Tunis.

In the neighbourhood of Constantine, the warm sulphur springs with baths, a section of the aqueduct built under Justinian, consisting of five arches standing in a garden of fruit trees, the new road around the mountain on the further side of the Roumel, and the gorge, are worth inspection. The last has been made accessible by a series of steps leading down to the bottom, along which runs a good path protected by an iron railing. On this one can walk through the gorge in about half an hour. The effect is marred by the discharge of several sewers from the city above, over the cliffs, which disfigures them, and is unpleasantly suggestive. High up on the rocks vultures and hawks build their nests and are constantly circling around in the air. In the gorge are two natural arches, beneath which the river pursues its course.

The Arabs, excepting the desert tribes, are cleanly in their habits. The lower classes bathe and change their linen at least once a week, the upper classes oftener. For bathing they use running or living water, wherever possible, and they are everywhere to be seen performing their daily ablutions in streams and at fountains. In respect to the adage, 'Cleanliness is next to godliness,' Christian nations can learn something from the Arab. In mingling with the multitude on fête and market days, we were surprised at the absence of unpleasant odours, so often noticeable in street-gatherings in more favoured lands.

The Jewish and Arab women of Constantine go out to the mineral springs to bathe and perform certain devotional rites every Wednesday. In our excursions around the city we saw men frequently sitting on the hillsides changing and mending their clothing.

The streets in the towns were almost universally well cared for, and we could not help contrasting the cleanliness of the towns with the filthy condition of many towns and villages in Italy, Sicily, Switzerland and South Germany.

During the fast of the Ramadan, which all Mohammedans in Algeria appear to keep religiously, the man who smokes or puts food into his mouth, between sunrise and sunset, is regarded by his fellows as a traitor to his religion. As an old historian puts it, 'No good Moslem will touch food so long as he is able to distinguish a black from a white thread.' In every town and city where there is a garrison, a sunset gun is fired during Ramadan, as a signal that the fast is over for the day. We had watched the scene that ensued in the squares of different towns with curiosity, but in none was it so pronounced as at Constantine.

As the sunset hour approached, the square near the theatre became filled with Arabs, who lounged, or walked about, or sat on the stone steps of the buildings in attitudes that only the Arab knows how to take. The vendors of bread, greasy batter cakes and galette, moved about showing their articles to the passive public, which was patiently awaiting the signal to eat. Others were engaged in frying fish at small portable stands, and as we passed, a lively sputter of fat in the pan was heard.

At last the gun was fired, when the crowd closed in upon the provision merchants, and in a twinkling, buyer and seller were busily engaged in devouring fish and grease cakes, as if eating was the one occupation of life. How the Arab fishman cooks his fish, eats himself, sells to dozens of hungry customers, and makes the right change all at the same time is a mystery, yet it is doubtful if he loses a sou. This scene, like the Arab market, is animated but never boisterous. Later on, when eating is over, liquor and absinthe have produced their physiological effect, and they are ensconced on the stone seats of their cafés drinking 'Café Arab,' their conviviality becomes sometimes disagreeably audible, as far as the European hotels.

CHAPTER IX

MEDRASEN – PIC DE CEDRES – TIMGAD, THE ALGERIAN POMPEII

A RIDE of two to three days southward from Constantine brings one to Biskra and the Sahara. The route leads over Ain Yacout, nine kilometres from which may be visited the highly-interesting Medrasen, or, as it is commonly called, the tomb of the Numidian kings. Some historians say it was built by the Massinissa family, others during the reign of Micipsa. The same slight haze seems to envelop the history of this monument, that floats in the air when precise facts as to the Numidians are sought after. The form is that of a truncated cone. It is composed of a series of twenty-four circular discs placed one above another, making steps fifty-eight centimetres high and ninety-seven wide. The upper disc has a diameter of eleven and a-half metres, and the lower of fifty-eight and a-half. A gallery was discovered in 1873, leading to a sepulchral chamber directly beneath the middle plate. The blocks of which the discs are made are large, and fit well together. It resembles 'le Tombeau de la Chretienne' near Kolia, which served as a tomb for a whole family of Moorish kings.

Batna, one hundred and eighteen kilometres from Constantine, is a place of little importance, except as a starting-point for the ruins of Lambesa and Timgad. A profitable excursion may also be made to the Tougourt or Pic de Cédres, a mountain about six thousand feet high, commanding an extended view of the Aures mountains and the distant Sahara, as well as towards Constantine. The last fifteen hundred feet is covered by a primeval growth of remarkably large cedars. This historic tree, which is dying out in Algeria, struck, as the Arabs say, by celestial malediction, is best seen here and at Teniet-el-Had.

The trip, to be thoroughly enjoyed, should be made in May or June; but if the traveller is hardy, and not afraid of exposure, the cedars may be visited in March, and very grand they then are, standing up tall and black, with huge, spreading branches, from out the snow which covers the mountain top. We climbed to a point about half an hour from the summit, when the snow became so deep we were obliged to turn back.

Tourists should be warned not to stop at the 'Maison Forestière,' where those taking carriages from the Batna hotels are deposited by the driver, with the remark, 'Here are the cedars; this is as far as we go.' The average traveller, whose faith is firm in the judgment of the hotel-keeper and his menials, looks with disgust at a dwarf cedar perhaps two feet high, gets into the carriage and returns the long distance to Batna, determined never to search after another cedar. But the sensible man leaves his carriage at the house where the road ceases, and ascends the mountain on foot for at least two hours, which must be done in order to see the trees in their full growth and glory.

A bright cold morning saw us in the saddle at 6.15, with the ruins of Thamugas or Timgad, forty kilometres distant, as our objective point. Our hostess thought the expediency of using our rovers for this expedition doubtful, said the country was wild and hardly safe, and that before reaching Timgad, two and a-half kilometres from the highway, a river had to be crossed, over which there was no bridge. The last statement had a somewhat dampening effect upon our ardour, but we started, notwithstanding the parting remark of the coachman, faithful vassel of the hotel, who called after us, 'Vous n'arriverez pas.'

Reaching Lambesa, eleven kilometres from Batna, in good time, we stopped to inspect the Roman ruins there, consisting of the usual amphitheatre, baths, basilica, temple of Minerva, etc. The open-air museum contains, besides the relics found at Lambesa, a number of the best statues from Timgad. A short distance beyond Lambesa, another town was passed with Roman ruins, the last on the route, which now wound over bleak hills, and finally, for several kilometres, through a high valley shut in on one side by the snowy Aures, and at the horizon by the Djebel Mettili.

The kilometre stones ceased, as did all signs of human life upon the road. Now and then a broken Roman column or crumbling arch was sighted, which afforded evidence that the Ancients found this a fair land to live in if the people of to-day did not. At length we saw some Arab shepherd boys, whom we questioned, but they only smiled in a daft way at the mention of Timgad. We began to fear we might have missed the route, when we came to a house inhabited by a Frenchman and his Arab wife. From the former, who happened to be intoxicated, after he had recovered from his surprise at seeing us, we managed to glean the information that Timgad was ahead.

Some distance further on we saw a post standing at the entrance of a cart-path, which led over a meadow. From this we learned that Thamugas, Roman city, lay two and a-half kilometres away. We turned into the path, which proved to be so rough that we were obliged to dismount and push our cycles. A shepherd here joined us, who spoke a few words of French, and we questioned him about crossing the oued or river. He said there was no bridge, and offered to carry us and the wheels over on his back for a certain sum. Finding the stream to be about a hundred feet wide, and the water of the ford about two feet deep, we accepted his offer. We were rather surprised when he gave the money paid him to a tall, sinister-looking fellow, apparently about thirty years old, who at that moment came up, and who said he was his son. The truth of this statement we doubted, knowing that an Arab father, who exercises parental authority over his children till long after they are of age, would be very unlikely to surrender his purse so easily to his son. The older man then went back to his sheep. The so-called son followed us, much to our annoyance, to Timgad.

Thamugas or Timgad, at first one of the military posts established to keep in order the turbulent pupulation of the Aures, rose to importance as a city about the year 100, under Trajan. It increased rapidly in wealth and population, was embellished with magnificent forum, temples, theatre, baths and other buildings, and received the name of 'Splendissima Civitas.' From the reign of Trajan to that of Constantine, it enjoyed great prosperity, and became the centre of a large district covered with flourishing towns and farms. In 305 began its decline, and its fortunes waned during the remainder of that century. In 429 it was sacked by the Vandals, who ravaged the country.

Shortly afterwards, Salomon, lieutenant of Belisarius, finding the city deserted and destroyed, built a fortress with columns and slabs of the forum, the steps of the theatre and other similar materials, which, though an act of demolition, served to preserve much of archæological value that might otherwise have been lost. For a time peace reigned in Thamugas. Later, the Christian Berbers occupied the city, but were so harassed by the Arabs, that in 692 they withdrew into the mountains. Timgad, having withstood several earthquakes, was destroyed for the last time, about 700, by a fire.

Only the central part of the city around the forum and public buildings, and a few of the principal streets, have as yet been excavated but what has been brought to light shows a luxury and completeness, in all the appointments for public convenience, that could scarcely have been surpassed in Rome itself. Future work will doubtless bring to light valuable material. No statues of the highest art, so far as we could learn, have been discovered, but those seen compare favourably with similar contemporaneous remains in other parts of the Roman Empire. The limits of the ancient city are obscured by the débris under which the ruins are buried; but that it covered a large area is evident from the broken columns which project up from the undulating surface of the soil for a long distance on every side of the excavated portion.

STREET AT TIMGAD.

Surrounded by these monuments of the past, which speak with a mute eloquence of the busy activities that were exercised here, the present fades from the mind and we are carried back to the time of Trajan and Constantine, when Timgad, splendissima with its noble buildings for men, and temples to the gods,

37

was the pride of an emperor. We live again with the people who laboured here and lavished their time and treasure in fashioning and adorning the city until it became an epitome of beauty and elegance. We picture their despair when the war-cry of the Vandal was heard as he and his hosts bore down upon, overwhelmed and ruthlessly destroyed it. From the ruins of this civilisation, the fire of whose hearthstones was quenched twelve hundred years ago, comes the question – for what purpose was all this endeavour? why this wanton destruction of so much of the beautiful and useful?

As we left Timgad on our return, we noticed we were shadowed by the same fellow who had followed us up, together with a companion, who had joined him. Nothing was said until the river was reached, when, no other Arabs being in sight, we addressed the one who had taken the money in the morning, and offered him the same amount then paid to put us again on the other side. This he was too cunning to accept. Taking advantage of the fact that we were on the farther bank of the river, he demanded double the price offered. We remonstrated, and, among other arguments, threatened to ford the river ourselves, on which the companion, pointing to his ankles, which were bruised and red, said the bottom was covered with sharp stones, which would wound our feet in a similar manner.

Argument availing nothing, we yielded to his demand, making it clearly understood that he was to transport us and all our belongings over. First he carried one of the party across. On his return, from unwillingness to trust him and the rovers alone together, he was ordered to take them next. This he now refused to do unless the exorbitant price already agreed upon was doubled. He was told that, unless he carried out his contract, his further services would be dispensed with and he would forfeit his reward. Evidently, encouraged by our acquiescence in his previous demand, and not believing the threat would be executed, he remained obstinate. Whereupon, the remaining member of our party drew off his boots and stockings, shouldered one of the wheels, waded easily across on the smoothest of sand, returned and took over the other, the Arab watching the procedure with some astonishment.

Both Arabs followed, and with a third, who now appeared on the scene, stood near while stockings and boots were again being drawn on. When this operation was completed, and we were ready to start, the man was offered one-half the price agreed upon, although he had proved false to his contract, and had performed only one quarter the work. He refused the money, insisting on the whole amount. He was told he must take that, which was more than he deserved, or nothing. At this he stepped forward, with an ugly expression, and laid hold of one of the wheels. This action, together with the discovery of a long, dangerous-looking knife at his girdle, and the recollection of the evil reputation enjoyed by the nomads of the Aures, made it evident that further trifling was useless. The muzzle of a revolver, within six feet of his face, accompanied by some forcible expressions, convinced him we were not so helpless as we had appeared. He took the offered money, turned without a word, and with his two companions forded the river and walked away towards Timgad.

The ride through the lonely region in the morning had seemed long, although accomplished in three and a-half hours. The return was destined to be longer. The highway, from the point where the cart-path joined it, ascended steadily for some fifteen kilometres. The grade was not so sharp but that, under ordinary circumstances, it could have been ridden, but a strong wind, which blew directly in our faces, rendered riding impossible, and we were obliged to make that fifteen kilometres on foot. This was not a pleasant complication in view of the experience we had just had, for we did not know at what moment the baffled Arab might bring the whole tribe of his wild companions, whose tents were visible across the valley, in pursuit. It would have added to our sense of security had we been able on this occasion, as on some others in Algeria, when in a disagreeable neighbourhood, to ride swiftly on and leave it behind.

The country was open and the road fully exposed to the sweep of the wind, and it was long past noon before we could find a place sufficiently protected to warrant stopping to satisfy our hunger. Finally, we drew up behind a clump of thorn bushes, which afforded a partial shelter from the cold blast, and opened our satchels. It is our custom, when en route for the whole day, to carry our luncheon with us. This secures entire independence, and saves time, which is of great importance on a long trip. It also assures a noon meal, which would often otherwise have to be dispensed with, since, in a sparsely-settled country, the uncertainty of arriving at mid-day in a town with an inn is very great.

Those who have not tried it do not realise the zest given to an unconventional meal in the open air by an appetite sharpened by a good morning's work. Sometimes one lunches beneath the soft skies of joyous spring days, amid grand scenery, by babbling crystal fountains, at other times overshadowed by threatening clouds, or perchance by the dusty roadside, under a burning sun, with no water to be had in miles.

Ten kilometres before Batna the wind ceased and a pouring rain began, which added the pleasure of a good soaking to the other adventures of the day. Arrived at the hotel, the landlady remarked, she had felt all day we ought to have taken a carriage to Timgad.

CHAPTER X

THREE OASES OF THE DESERT — EL KANTARA, BISKRA, SIDI OKBA

WE set out from Batna, on the ride to the desert, with some misgivings, as we had been unable to obtain accurate information, even from the bicycle club in Algiers, as to whether the road ran as far as Biskra. Some thought it did, but the man who knew positively, or who had been over it, was not to be found. The road to El Kantara, fifty-sixty kilometres, was very fair. Beyond the station it passes over a modernised Roman bridge, spanning, with a single arch, the river that flows through the gorge cleft in the high and rocky mountain wall, which gorge the Arabs have for centuries called the mouth of the Sahara.

From the bridge is unfolded a striking coup d'œil; behind lies a barren waste of mountains, hills and sandy sweeps; in front, at the opening of the gorge, twenty thousand palms lift their tufted tops toward the blue sky; beyond these is the desert. One sees for the first time the Sahara, and, with the same glance, an oasis of great beauty and verdure.

After leaving El Kantara, the road became poor and stony, degenerating at last into a caravan track, which ran up and down among the hills, and over shallow streams, sometimes so spread out that the right direction was difficult to find, and sometimes ceasing entirely. Arab horsemen and lines of camels were met with, coming from Biskra and Tougourt. It is a pity that roads cease in the land where the camel reigns, for he takes the presence of bicyclists, and especially of women bicyclists, much more calmly than either European animals or people. He neither accosts a woman with the trite remark, 'Das ist nicht schön,' as the German often does, nor does he run up a sand-hill to escape from her presence. But he bears the unusual sight with composure, and, with only a gentle look of surprise in his large soft eyes, he pursues his line of march, not varying an inch because two bicyclists happen to invade his domain.

Our frequent meetings with this quiet animal, together with the bad condition of the road, made us realise that we had reached a land where the Arab and his camel could probably travel with greater ease than the American with his wheel. It was odd work exploring the way to Biskra, but we managed somehow to come out on the road wherever it existed.

After a long climb over a mountain pass, the road, which here was well made, turned abruptly around a spur, bringing Biskra, six kilometres distant on the plain below, into full sight. In 1842, the Duc d'Aumale and his army, having lost their way, after several days of aimless wandering, came upon this spot, the Col de Sfa. The weary soldiers sank upon the ground, exclaiming with joy, 'La mer! la mer!' – a mistake not surprising under the circumstances, in view of the resemblance to the sea which the desert here presents.

The mountain upon which we stood culminated in a sharp point several hundred feet higher, crowned by an old Turkish fort. No path led to it from that side, so, hiding the wheels, we climbed up the rocks, feeling certain the view from above would repay the exertion. After a scramble of half an hour, with some scratches from rocks and occasional thistles, we reached the summit.

We had met with barrenness and lack of vegetation before in Algeria, but the desert motif, as seen from the old Turkish fort, was for the first time complete. Vast and mighty as the ocean the Sahara lay before us, stretching outward until lost in limitless haze. Besides the oasis of Biskra, comparatively in the foreground, a number of others were visible, the dark colour of which, contrasting with the lighter hues of the plain, gave the leopard-skin appearance which has been mentioned in connection with this view. At various points, clouds of dust raised by the wind filled the air, into one of which a caravan starting out from Biskra disappeared.

As we rode down toward the sand-hills that lay between us and Biskra, we realised the beauty of its situation; a vast palm garden in itself, with the desert on one side and the mountains on the other, purpling in exquisite sunset tints as evening approached.

Since the opening of the railway, this town has become somewhat of a health station. Cook makes it patent to his patrons that Biskra can be reached from Algiers in two days, with a stop of one night en route at a comfortable hotel, where a palatable table-d'hôte is served on arrival of the train, so that travellers are enabled, without especial hardship, to get a glimpse of the desert, or go to Biskra in search of health. Cook omits to mention in his list of inducements that the carriages are not any too convenient, being antiquated in style, and having the appearance of having been cast off by the French railways, and that the low rate of speed of Algerian trains is only equalled by those of Spain. Later on, being obliged to take the rail where there were no highways, we had an opportunity of experiencing some of the disadvantages.

The hotels of Biskra are neither very bad nor very good, but the days when the romantic German tourist sat alone in the Moorish dining-room of the Hotel de Sahara, served by silent-footed, sad-eyed Biskris in national costume, are unhappily at an end. At the present time, the crowded restaurant, the French menu and the clatter of the table-d'hôte are found, and although it is the Biskri who serves, in native trousers and fez, he wears a French sack, and is very French in manner.

The houses of French Biskra, like those of other Ziban towns, are built of sun-dried bricks made from a sticky clay of this region mixed with chopped straw, are low, of at most two storeys, and generally placed over arcades.

STREET IN VIEILLE BISKRA.

Two kilometres south of the modern town lies Vieille Biskra, a rambling old village of mud houses built around open courtyards. Nothing could be more picturesque of the kind than the long irregular streets bounded by mud walls, some with canals through the middle, which intersect it in all directions. The town extends, under different names, for five kilometres along the Oued Biskra, which partly supplies the water necessary to the existence of the oasis, beyond which it soon loses itself in the desert. Everywhere, around the houses, overhanging the street walls, covering all the open spaces, rise the date-palms, not like the dwarfed and stunted specimens seen on the French and Italian Riviera, but with luxuriant growth and bold freedom shooting up forty to fifty feet in the air, as if they revelled in existence. The number of palms in Biskra is said to reach a hundred and forty thousand. There are also some six thousand olive trees, and one superb cypress, which contrast effectively in form and colour with the palms.

The Biskri inhabitants of this village gain a livelihood by cultivating the palm, and, to some extent, the cereals. Judging by the character of their houses, and the almost entire absence of furniture, their wants must be slight. Possessing nothing of value, they are said to bear philosophically the occasional disappearance of their houses under the softening action of winter rains, and to go quietly to work reconstructing them as soon as the dry season opens. Every palm is taxed by the Government a fixed sum, so that the owner can easily calculate the amount of his tax.

Just outside modern Biskra is the negro village, inhabited entirely by Ethiopians, who chatter, dispute and beg of strangers in a manner unknown to the quieter people of Vieille Biskra.

The modern town is a dull place, and is over-full of guides and boot-blacks, who spring up at every step. There is a trifle more excuse for the boot-black than for the guide, for after a rain – and it is often after a rain in Biskra – no matter how shiny one's boots are on going out, a walk of five minutes covers them with red mud.

No visit to Biskra would be complete without an evening spent at the café houses of the street of the Oulad Naïls. The young women called by this name come from the Oulad Naïl mountains, south-west of Biskra, to earn their dot as dancing-girls in the Arab cafés. The guide-book warns travellers not to take a guide to the cafés, as they will get off cheaper without this encumbrance, advice not usually heeded, as the train arrives at Biskra in the evening, and after table-d'hôte the newly-arrived visitor's first thought is to visit the cafés. To avoid groping about in the poorly-lighted streets, he is almost certain to take one of the officious guides loafing about the hotel entrance. He is soon seated on one of the narrow benches of the 'Café Maure,' facing a group of Arabs curled against the opposite wall smoking and sipping coffee. He orders coffee or syrup and treats the guide, his ears, meanwhile, being tortured by most barbarous music, and his eyes blinded by the smoke of tobacco. The guide nudges him. 'Elle va venir maintenant,' he says, and as he speaks, a girl in bright drapery, with long horse-hair braids, painted eyelids, and ornaments on her neck, arms and ankles, walks slowly through the café and out of the opposite door. 'Why does she not dance?' the visitor asks. 'Oh,' the guide says, while he sips a second café, 'there are not people enough here, but if monsieur is willing to pay two francs.' Not caring to bear the burden of expense alone, the visitor goes to another café, where coffee is again ordered, and a like experience met with; or perhaps, if other strangers are present, one of the girls condescends to dance for a few minutes.

But they are much too spoiled to dance long, and the visitor soon departs disgusted, not without a suspicion that he has afforded quite as much amusement to the Arabs as they have to him. The street scene outside is more interesting. The native population, French soldiers, zouaves and strangers from every land, make this short street their evening promenade between visits to the cafés. The Oulad Naïls, in their best garb, and profusely decorated with ornaments, move about singly, or two together, or sit on the doorsteps of their houses, attracting much attention, especially when a bright moonlight adds softness to their Oriental charms.

We noticed, in our promenade in this crowded street, that if anyone brushed against us it was always a European. The Arab has a natural way of passing so closely that his presence is felt, yet without touching the person met. He seems to shrink into his drapery, or turns half aside with an innate grace that never fails him in a crowd. Here, as elsewhere, in studying the people, we could not help admiring the erect, finely-developed forms, the natural, easy attitudes, and the untrammelled freedom of movement peculiar to the Algerians, to which the people of civilised nations show no approach.

Among the attractions of the Biskra market are embroidered leather bags and purses, toilet bottles and hand mirrors, covered with bright-coloured embroidered leather adorned with yellow, green and scarlet tufts of wool such as are used by the Oulad Naïl women, and knives with bone handles and red leather sheaths, which are worn by the men at the belt. Perhaps the most original souvenirs are the stuffed palm lizard, about a foot long and of a grey colour, and the desert or sand lizard, which is much larger, with a long, tapering tail.

A short distance west of the town a low rocky hill rises out of the plain, commanding an unobstructed view in all directions. No lover of nature should fail to visit this point at sunset on fine evenings, as the colour phenomena are such as cannot be seen every day nor everywhere. Particularly noticeable are the Aures mountains to the east, the grey, naked ledges of which, catching the waning light, take on from base to summit a fascinating combination of mysterious greens and pinks that charms the eye, and can be compared only to the fire and glow of a fine opal.

At Biskra an excellent opportunity is afforded for seeing camels, which are constantly arriving and departing. Light and dark-coloured camels, well kept ones with soft handsome coat, and neglected ones with skin denuded of hair, and other marks of ill-treatment; some with good and some with most meagre trappings, may be found at any time at a place just outside the town.

As we were photographing some of the different specimens, we came to a small mud house on the outskirts, in front of which a camel was resting. While making preparations to photograph him, his owner, a desert negro, evidently thinking we were about to practise some incantation on the animal, came running out with cries to protect his property. As soon as he came within proper range, the Kodak was snapped. His presence in the resulting picture, with one foot in the air, furnishes an example not only of an object in motion, but of what might be termed a photograph of a mental state.

One beautiful sunny morning we started from Biskra, which is the second oasis, to ride twenty-two kilometres to the third oasis, called Sidi Okba. After about a mile we came to the Oued Biskra, which is quite wide at this point, and over which there is no bridge. Here we waited for some one to come up who would be willing to take us over. At length a strong, well-formed Arab appeared, driving some mules. We had no sooner made known our wishes than he offered to perform the desired office. Taking us pick-a-back on his shoulders, he carried us to the opposite bank. On other similar occasions some Arabs preferred this method, others took us up in their arms as one picks up a child.

The road to Sidi Okba was a natural clay or mud road, having a ditch on both sides for drainage. The heavy rain three days before had softened it so that the wagon wheels and camel's feet left ruts and tracks over a foot deep, and in some places had washed it away bodily for fifty feet at a time. Riding on that road, in such a condition, was an impossibility, so we were obliged to take to the desert, which here was covered with patches three to ten feet in diameter, and perhaps a foot high, of a brittle green plant, between which a skilful rider could pick his way. We were obliged to dismount frequently and walk around, or through small gullies washed out by the water, so that progress was slow.

After a while the disagreeable March wind sprung up, hindering us still more, and driving the sand into our faces, so that our cheeks tingled. We met caravans of camels, and Arabs on horseback with burnous hoods over their heads to protect them from the sand. We wondered if it would be our fate to sit for ten hours with our backs to the mistral, as did the Duc d'Aumale on this same stretch with his to the Sirocco.

One likes to think of Sidi Okba, the religious capital of the Ziban, containing the oldest mosque in Algeria, as an interesting place, but with the best intention it is difficult to get beyond the French expression, bien curieux, in regard to this miserable hamlet with its poverty-stricken inhabitants, many of whom are affected with eye diseases. One even feels indifferent to the important fact that the bones of the warlike marabout, Sidi Okba, are covered by the sands that support the mosque. This cherished and venerable Moslem has a way of cropping out so constantly in the history of Algeria, that it is with a certain sense of relief that one learns that he finally succumbed to the Berbers at Tehouda. Arabs, from religious fervour, may care to repeat the pilgrimage to the portals of Sidi Okba, but one visit will satisfy the tourist.

CHAPTER XI
INTO THE SILENT LAND OF ROMAN RUINS AND ARAB GOURBIS – AIN BEIDA – TEBESSA

WE found the last ten days of March and first half of April more rainy than February and the first half of March had been. This is said usually not to be the case. Haeckel, the German scientist, who visited the country a few years ago, had an experience similar to ours, and was prevented by rain from carrying out his trip as he had intended.

After several rainy days we left Khroubs for Tebessa, via Guelma, where we expected to spend that night. The sun shone in the early morning, but people shook their heads dubiously when asked if there was a chance of the next two days remaining fine. Good weather was of especial importance to us just then, with a two days' ride of over two hundred kilometres, through a sparsely-inhabited country, on our hands. After a run of fifty-six kilometres, of which thirty were steadily up, we reached Oued Zenati, a small town in a basin between high hills, before eleven o'clock.

While buying our luncheon, knowing that at noon we should be miles from a town, someone asked where we were going. On naming the place, they told us that the road indicated on our map did not exist, and that to reach Tebessa, we must cross the mountains in another direction and pass the night at Ain Beida. This was rather astounding news, for we knew nothing of the town just mentioned, and had previously been informed that the intended route was practicable. We studied the faces of those gathered about, wondering whether they were honest, or, indeed, knew anything about the matter.

On questioning the best-looking man, a post driver, as to the distance to Ain Beida, he replied, 'Seventy-five kilometres.' We glanced at our watches. It was eleven o'clock. Were there any towns on the way? Was the road hilly? No; there was an Arab settlement on a high plateau that had to be crossed, and later, a post station at the top of a pass. Clearly, the road was both lonely and hilly, but the idea of stopping where we were until the next day did not appeal to us, and we decided to try the new route.

The sun was still shining, but great masses of cumuli were chasing one another over the sky, which we feared might at any moment blow together and form rain clouds. Although we felt it was a useless thing to do, we appealed to the post-driver for the last time. Would it rain? As we expected, we received for an answer the usual French shrug. Never once in the whole journey did we meet with a man who would venture an opinion on the weather, and our experience led us to the conclusion that the reserve manifested in regard to this topic was justifiable.

We started, and from that moment on, until Tebessa was reached the following afternoon, dread of rain haunted our minds more than fear of brigands had ever done. That morning there had been a frost and it was still cold. We knew the capacity of African mud for miring our wheels, and if mired on the uninhabited plains ahead, we should have to spend the night in the open air in a freezing temperature.

We got along well enough for about twenty kilometres, climbed some hills and came out on an extensive plain. The sky looked decidedly threatening. On one side, not very far off, it was raining, and in front, a huge, black cloud was rapidly approaching. The bulk of it passed away to the right, but enough swept over us to give us the benefit of a hard hail-storm for ten minutes. Some Arabs, with hay-carts, meeting us at this moment, stopped to cover their horses, and we and they sought shelter behind the wagons. After this the sky looked more threatening than before, and by the time we reached the miserable Arab hamlet, the clouds in front were as black as those behind, and it appeared to be raining not far ahead. The whole country seemed flooded with water. Fortunately, the road was well made and hard. On the plain we came to four or five places where the water from the meadows was flowing over the road in streams six to ten feet wide.

As the water was turbid, we could not discover how deep these little rivers might be, but they had to be passed; so, taking the chances of being upset or caught in a quicksand, we put on extra speed and plunged through, escaping with no damage except wet feet.

We now ascended again over another mountain range, where it rained again, wetting the road considerably. At the house at the top of the pass we drank some water and hurried down into another plain. There was no more rain. Now and then we passed Arab gourbi encampments at a little distance from the road, and one good-sized house – the only one in all the region. Arabs were sitting about in front of it, and a number of handsome horses, saddled, were eating grass in a field close by. The house doubtless belonged to some Spahis or wealthy native.

In the central and western parts of Algeria, outside the towns and cities, the rural Arab population live in villages, with thatch-work houses surrounded by a strong hedge of thornbush or prickly pear to keep off dogs and intruders. In the Province of Constantine the people are more nomadic in their habits, and live in tents or gourbis made of white or light-coloured canvas. As the desert is approached, the dark-striped gourbi of the Bedouin is seen.

The sun was setting and there was no sign of a town ahead. It grew dark, and still no sign. Fortunately, the clouds broke and the sky became clear, as it does suddenly in Algeria, disclosing the rising full moon, which lighted the way. We blessed that moon, but still more the faint lights of Ain Beida, which came in sight about seven o'clock. Just as we entered the town the road became very wet and muddy. We afterwards learned it had rained there for three hours that afternoon.

The best inn of the place was one of the most uncomfortable we met with in Algeria. The small, cell-like bedrooms, with stone walls and stone floors, placed directly on the ground, opened out of a narrow courtyard. They had no fireplaces, and, with the temperature outside near the freezing point, were bitterly cold. We asked if there was no higher floor, and were informed that no house in Ain Beida had more than one storey. The simple beds, provided with the coarsest of sheets, and two chairs formed the furniture of the room. After considerable conversation with the chambermaid, an old French peasant woman, a scanty supply of toilet conveniences was produced. We decorated the wall with our pocket thermometer, and after a day's ride of one hundred and twenty-seven kilometres, of which seventy-five were up hill, slept the sleep of tired travellers in a temperature of 44°.

The peasant woman brought in two glasses of coffee at six in the morning, and we were soon out of our uncomfortable quarters. It was sunny, and the day augured well, but the roads were swimming. Nothing would have induced us to stay another hour in Ain Beida, so we started, expecting to be stalled in the mud. But mud and sun were better than the deadly atmosphere of the inn we had left.

Riding the road was out of the question, but we walked on the grass at the side for five kilometres, when the wet zone ceased and the road became dry again.

Around Ain Beida, within a radius of forty-five kilometres, are scattered columns and other Roman remains. At Ain Gueber two Christian inscriptions of the third or fourth century have been found. One of these reads – 'Spes in Domino et Christo ejus.' Cherbonneau, speaking of this, says, 'Those who visit that desolate region to-day are not content with placing their trust in God only, but provide themselves with revolvers as well.'

The distance to Tebessa was ninety kilometres, and two mountain ranges had to be crossed. There was but one village on the route, and this lay in the valley between the mountains. The sky clouded over before we reached the town, and by the time we arrived there appeared about as dubious as on the preceding day. From this place to Tebessa was the longest distance ridden during our trip between two towns – viz., fifty-four kilometres, with but one house on the way.

As we were ascending the last mountain slope rain fell in large drops from time to time. We did not dare stop for luncheon, but ate as we walked. Here an encampment of Bedouins was passed. Their gourbis were brown striped with red. Men, women and children came out of the tents and followed us for a while, but said nothing. They were scantily clothed, and some of the women wore horse hair in their braids, and had tattooed faces. They resembled the Bedouins of the desert, which they doubtless were, as this region is not far removed from the Tunisian Sahara. We met the diligence tightly booted, as if expecting rain.

Showers greeted us as we rode down on the other side, but were not minded, as the road on the mountain was hard and good. When, however, we came out of the Plaine de Chabro, on which Tebessa is situated, about twenty-five kilometres from the city, our fears were partially realised. As is usual in Algeria, the road on the plain was not as well built as in the mountains, and the rain had put it in very bad condition.

At short intervals the wheels became so clogged with mud that we were obliged to dismount and clean them with sticks, which operation took a good deal of time. In two places the street was so cut up that we had to carry the cycles a hundred yards or more. We were beginning to feel anxious lest we should be kept out over night, when the muddy, rutty road suddenly changed to a well-built one. In another half-hour Tebessa was in sight at no great distance.

How much the chance of improvement in a road raises a bicyclist's spirits in touring in foreign lands! To be sure, the probability of its becoming worse is just as great, but the wheel enthusiast is an optimistic fellow, who does not harbour such probabilities in his thoughts. Thus ended the mental tension of the two days' trip through a deserted country, occupied mostly by nomad tribes, and frowned upon by the most fickle of skies. With a sense of relief, we rode into Tebessa.

The modern Tebessa, seventeen kilometres from the Tunisian frontier, stands on the site of the once important Roman city of Theveste, at the entrance of the valleys that lead into the Tunisian Sahara. It is also accessible to the valley of

Kairouan. The neighbouring mountains are covered with pine forests, and contain marble quarries. When the railway which connects the town with Souk Ahras has been carried out to Gabes in Tunis, which will shortly be done, the French expect the Tebessa of the future will be as flourishing as was the Theveste of the olden time.

The types here are varied, and the burnous worn is often brown, dark red or orange, which adds notes of colour to the ordinary white drapery. The Arab, Tunisian and Negro seem to have especially well-developed physiques, and afford splendid specimens of Oriental humanity.

The history of Theveste is similar to that of Thamugas. Founded about 72 A. D., it was raised to the position of a Roman colony under one of the Antonines. It reached the height of its prosperity in the third century, under Septimius Severus, when its best monuments were built. Like Timgad, it was destroyed by the Vandals, and partially restored by Salomon, successor of Belisarius, in 543.

Of the Roman relics, the most beautiful is the quadrifons triumphal arch erected in 212 to Severus, Julia Domna his wife and Caracalla their son. It now serves as one of the gates of entrance to the town, and in its nearly perfect state of preservation is worth a long journey to see. The natives have a way of loafing lazily under its arches, and were it not for the Arab features, one might imagine the Roman, in his toga, leaning against the stones he has so effectively reared.

Near by stands the well-preserved temple of Minerva, a highly-ornamented Corinthian structure, resembling, in some respects, the Maison Carrée at Nimes. Outside lie statues and tablets found about Tebessa. The inside is being fitted up as a museum.

At a short distance from the city walls is a large area covered with ruins, forming a little city in itself. These ruins are called the Monastery, which was built about the beginning of the fifth century on the remains of a heathen basilica. One of the most interesting and best-preserved parts are the stables, where the stalls, each with its stone trough, and hole through the post for tying the horses, remain exactly as when made.

A few kilometres south-west of Tebessa, near some ruins, tracks in the rock made by the wheels of Roman wagons may be seen for a distance of two kilometres. From the nature of the tracks, archæologists say the road was used for conveying marble from the quarries in the vicinity, which were known and worked by the Romans. Samples of the fine red marble have been placed in the museum at Constantine. In ten different places, within forty kilometres of Tebessa, Roman ruins are to be found.

CHAPTER XII

TUNIS – CARTHAGE

NO carriage road existing between Tebessa and Souk Ahras, we were obliged to take the rail to the latter place. The train left Tebessa at six o'clock in the morning, and arrived at its destination about noon, moving with the astonishing average velocity of twenty-one kilometres, or thirteen miles, per hour. The temperature on starting, and for three hours afterwards, was not far from 40° F., and the only first-class coupé on the train was unheated. It may be imagined that sitting for six hours in such a temperature, in light riding costume, without over-garments or wraps of any kind, was not promotive of bodily comfort.

Souk Ahras is the last Algerian town of any size on the railway line between Constantine and Tunis. Between Souk Ahras and Tunis there is no highway, so we left the wheels in charge of the stationmaster, and took the afternoon train for Tunis. The distance of two hundred and forty-seven kilometres was accomplished in eight and a-half hours. The rate of speed on this line was a little higher than on the other, being twenty-eight kilometres, or seventeen and a-half miles, per hour. On account of the tedious railway journey, Tunis can be more conveniently reached direct from Marseilles or from Sicily.

Tunis has no natural advantages of situation, being built upon a sloping plain, but it is well worth visiting on account of the graphic pictures it affords of Oriental life, not having been affected so much as some other cities by European influences. The people are very different from the Algerians, being more rotund in figure, apparently better fed and more richly dressed. The men have a sleek, well-nourished, oily appearance, that betokens a luxurious manner of living, and present a striking contrast to the long-faced, angular-featured, hungry-looking Arabs of Algeria.

The women, particularly the Jewish women, are enormously stout. Their huge, unwieldy legs below the knee are swathed with bands, between the folds of which the skin presses out. Girls are married at the age of ten or eleven years, previous to which ceremony they are subjected to a stuffing process, to fatten them to the standard required by custom for entering the matrimonial state, and they do not appear to lose flesh as they grow older. The women may be seen in the streets walking slip-shod in short slippers, with the heel under the instep, which gives them an ungainly gait. The rich wear a white silk burnous with a pointed hood. Many paint a black line across the forehead at the level of the eyebrows.

The people are enterprising, industrious and seemingly prosperous, and are occupied during the day with some business or trade. The evening is given up to amusement. Then the streets of the native quarter are thronged with people; hawkers of various wares shout from their portable stands; carrousels, side shows of different kinds, and cafés chantants are in full operation.

Perhaps the greatest curiosity of Tunis is the Bazaar, or Souk, a large area of narrow covered streets and passages, out of which open small boutiques six to ten feet square, and some much larger, where are sold all kinds of articles. In the bazaar of the perfumers are found essence of rose, jasmine, geranium and other perfumes, with which, in the presence of the purchaser, the well-known long and narrow glass flasks with imperfectly-fitting stoppers are filled. The lady who provides herself with a goodly supply of these to take home to her friends is likely, in a day or two, to find the contents transferred from the flasks to her travelling effects, with the result of disgusting her for ever afterwards with the odour of attar of rose and essence of geranium.

The bazaars of the tailors, the shoemakers, the hatmakers and the jewellers all have their attractions, but that of the embroiderers is the most distinctive, and perhaps best worth attention. This class of merchants appears to be of higher station than the others, have larger shops, and carry a larger line of goods. When a purchaser enters they receive him politely, offer a chair, and serve a cup of Arab coffee, then proceed to business. The elaborate designs and rich execution of Tunisian embroideries are too well-known to need description, as are the softness and beauty of some of the silk textures.

The merchant is a wily fellow, and an adept in judging the pulse of his customer's inclinations. He states the price with decision, and the assurance that it is very low, bears with composure any amount of argument, knows when to yield, and will ultimately accept two-thirds, or, in some cases, one-half the price stated, rather than lose a bargain. Notwithstanding his previous assertion that the goods would be thrown away at the price offered, he delivers them, at the close of the transaction, with an air which implies satisfaction with his percentage of profit.

The road to the Bardo passes under one of the arches of a well-preserved section of the Roman aqueduct which formerly supplied Tunis with water.

The palace of the Bardo, while showing fine points of Moorish architecture, among which is the staircase of the Lyons, has been allowed to fall partly into a ruinous condition, and conveys the impression that the present Bey of Tunis has no desire to keep his property in repair. The museum was closed to the public at the time of our visit.

The name Carthage recalled the enthusiasm with which our youthful minds were inspired by the story of the Punic wars during the period of our classical studies, and the proximity of the 'ruins' of this historic city determined us to give half a day to inspecting them. After walking over the wide area said to have been covered by Carthage, but now covered with grass, and seeing nothing except some Roman cisterns, a few foundations, and occasionally a shapeless mass of stone and mortar, we came to the conclusion that Carthage was destroyed so thoroughly, both by the Romans and Vandals, that nothing remains now to interest the traveller, unless he be possessed of an extraordinary imagination.

It seems incredible that a vast city which flourished for six hundred years, during which time it dominated the northern coast of Africa, Sicily, Sardinia, the Balearic Isles and a part of Spain, and again, after its reconstruction under Cæsar, was for four hundred years the second city of the Roman empire, could have been so completely annihilated that only here and there an obscure trace remains.

CHAPTER XIII
THE HOT SPRINGS OF HAMMAN MESKOUTINE – FROM SETIF TO BOUGIE, THROUGH THE GORGE OF DEATH

FROM Souk Ahras began the journey westward. Passing through Guelma we came to Hamman Meskoutine, renowned for its hot springs. Algeria possesses one hundred and seventy-three hot and mineral springs, of which a hundred and twelve are in the Province of Constantine. Of these, the springs of Hamman Meskoutine are the most noted. They issue from the earth at various points over a length of two kilometres. The water, when it emerges, has a temperature of 203° F., and is heavily charged with lime salts, which are abundantly deposited on cooling. The deposit around the principal springs has formed a considerable hill, perhaps fifty feet high, that has been compared to a petrified waterfall or cascade, which it really resembles. Its surface is beautifully coloured in red, yellow and green. The water from the springs above, flowing over the slope of the hill, is carried off by a stream at the base, the colour of which is bottle green. The largest spring discharges about twenty-five thousand gallons per minute. Connected with the springs is the inevitable bath and water-cure establishment, well patronised by sufferers from rheumatic affections.

A delightful vista of rolling hills, upon which grow the almond, eucalyptus and olive, forms a graceful setting to the springs. Connected with the hotel and bathing establishment are attractive grounds, ornamented with Roman relics found in the neighbourhood – an indication that the Romans also availed themselves of the properties of these springs.

From Hamman Meskoutine we rode in two days to Setif, the starting-point of the road through the gorge of Chabet-el-Akhra to Bougie. This route passes through some of the most remarkable mountain scenery in Algeria, which, while quite different in character, can be compared in grandeur only with points in the Grande Kabylie. The distance, one hundred and thirteen kilometres, can be traversed in a day on a bicycle, but requires two days in a carriage. The traveller, therefore, who is loath to leave the railroad seldom journeys over this route.

The trip is more conveniently made from Setif than from Bougie, for while there are ups and downs in both directions, the general tendency from Setif is downward, the fall in the one hundred and thirteen kilometres being over three thousand five hundred feet to Bougie, at the sea-level.

The first part of the way lies through a barren, deserted mountain region, where no trees are seen, and fields are fallow. Here we met no people except an occasional shepherd tending his sheep and amusing himself by playing strange, wild tunes on a reed pipe. As we swept by the music would cease, and when we glanced back, we could see the shepherd standing immovable by his sheep, the reed half-way to his lips, his gaze fixed upon us as if wondering what manner of creatures were flying through the land.

At the top of an ascent about half way we stopped at a small village called Takitount, which has an exquisite view of the circle of the Chabet-el-Akhra mountains. At the inn, an excellent table water was obtained, similar to seltzer. It is bottled at a spring near by and sold all through the province under the name of Eau de Takitount. Several kilometres further, and much lower, just at the entrance of the gorge, is the village of Kerrata, with two hundred and sixteen inhabitants, which has an important Arab market on Tuesdays, drawing Arabs from the Chabet mountains, and those of the Petite Kabylie.

The market was just breaking up as we rode in. While we were taking black coffee on the inn veranda, the French patron was beset by scores of natives, with questions as to how we mounted and how we propelled the machines, although none were so dull as the Sicilian official who, in a town behind Mount Aetna, asked if we sat on the handle bar. That their chief curiosity was in regard to the woman we inferred from the frequent repetition of the word mujer, which may mean woman in their dialect as well as in Spanish.

Just beyond is the entrance to the Chabet-el-Akhra, or gorge of death, which winds for six kilometres between rocky, precipitous mountains, the sharp peaks of which rise with dolomite-like abruptness to a height of five and six thousand feet. Wild crags overhang the road, and in some places the opposite cliffs come so near together that only room is left for the road and the roaring torrent that rushes and plunges three hundred feet below, and the sun's rays penetrate only for a few moments at noon. Here and there side gorges open, disclosing vistas of other mountains quite as wild and grand as those through which the road runs. The road is a wonderful piece of engineering skill, built on the sides of the mountains with a bed of solid rock, and at times supported on arches. It is as smooth and hard as a floor, and perfectly cared for.

On leaving the gorge, the scene changes as if by enchantment. The bleak uplands crowned with treeless summits, and the sombre grandeur of the gorge, are left behind, and we come out, still skirting the bases of bold mountains, into the most fertile of valleys covered with beautiful forests of cork oak, white poplar, olive and mastic trees, while along the roadside bloom myrtle, laurel and various wild vines, the whole forming a semi-tropical paradise.

To add to the beauty of this Arcadia, a turn in the route brought the Mediterraneam into full sight, illumined by a golden sunset. We rode along the famous coast overshadowed by the castellated heights of Beni Tizi, Bou Andas and the bold summit of Babor, the highest mountain of the Petite Kabylie, and so lovely was it all, that, before we were aware, night began to close in, and we realised we could not reach Bougie by dark.

Coming to a small town, we asked, as bicyclists sometimes have to do, the first good-looking man we met, if there was a public-house fit to pass the night in. He said there was one, but advised our returning three kilometres to a place called the 'Rendez vous de Chasse,' where we should be better cared for. We followed his advice, and in about ten minutes reached what we had supposed, in riding past, to be a private villa surrounded by orange groves. Near by in the garden was a smaller house, in front of which were chairs and tables. We asked here of an old woman if we could be accommodated for the night.

While talking, an individual approached us from the villa. Evidently a woman, she was not attired in the costume commonly worn by her sex, but in a complete man's suit of tightly-fitting white jean trousers, high gaiters and short sack coat. She asked what she could do for us, smoking the while a cigarette. While a room in the villa was being prepared for us, she invited us to inspect the orange and lemon orchard. Her chief interest seemed to be in the cultivation of these fruits, and in her favourite sport of hunting. We afterwards heard her spoken of as an excellent shot.

Our lodging in the villa proved very comfortable, and the attendance good, except that the servant polished one pair of yellow boots with blacking. When we went over to dinner at the small house, we were taken through the large front room, used as restaurant, to a pretty little dining-room furnished with old carved chairs and buffet. A young girl, who might have been a sister of the hostess, dressed in woman's garb, with a garland of fresh flowers twined in her hair, served a dinner of six courses, in which fowl chiefly appeared under different guises.

As we went through the restaurant again, we saw Madame sitting, in the same costume, at a large round table serving soup to a dozen people, evidently dependants and servants of the house, who sat about her. A cheerful fire burned on the hearth. Having to make an early start in the morning, we bade her good-night, and accompanied by the pretty maid with her chaplet of flowers, carrying lights and sirop de grenadine, returned to the villa. We were soon asleep, dreaming of orange-gardens presided over by cordial amazons in trousers.

CHAPTER XIV
BOUGIE – THE INN OF TASMALT – THE RACE WITH THE CAVALIER – ARESKI, THE KABYLE FRA DIAVOLO

BOUGIE, a dull little French town with bad hotels, has a magnificent situation, and its surroundings offer some charming excursions. It is built on the rocky flank of Mount Gouraia, overhanging the Gulf of Bougie, and as it runs backward, is lost in pomegranate, fig and orange orchards, while facing it to the east, across the bay, the beautiful amphitheatre of the Chabet-el-Akhra and Petite Kabylie peaks rise in varied outline. Naples, Algiers, and even Palermo, are forgotten as one stands on the terrace at Bougie, and it is remarkable that so few writers, besides Maupassant, have praised its site.

It had been our intention, ever since entering Algeria, to visit the Kabylie. This region, though lying within a day's ride of Algiers, seemed to be a terra incognita, about which we were unable to obtain any very definite information.

When in Algiers, we asked our French official friend about it. He had been there, and advised us by all means to go; and as to the time, said, 'In Algeria, one can go anywhere in March.' Beyond this we gathered no information from him. A day or two later he mentioned the subject again, and warned us not to go, as his wife, who evidently exercised considerable influence over his opinions, had informed him, on his return home, of the capture of forty-two brigands, and insisted upon it that the country was not safe. We might have replied that the fact of the capture of the brigands afforded all the more assurance of security.

From the bicycle club we learned that there were two ways of entering the Kabylie – one from Algiers to Tizi Ouzou and thence to Fort National, the other from Tasmalt on the south-east side of the Djurjura mountains over the Col de Tirourda pass, seventeen hundred and sixty metres high, within five hundred feet of the altitude of Mount Washington. No one, however, could tell whether the higher roads were sufficiently free from snow to permit the carrying out of the projected tour. In the uncertainty we finally decided, as it was still early in March, to wait a month, until our return from the desert, and then try to enter over the Col de Tirourda. Tasmalt was, therefore, our next objective point from Bougie.

As the guide-book mentioned no inn at Tasmalt, where it would be necessary to pass the night, we had written ten days before to the Chef de Gare at that place, asking if there was one, but had received no reply. At the Bougie hotels we were told one existed, but could not learn its name. As a last resort, we applied to the post-master at Bougie. He was very friendly, but, as was to be expected, knew no more than the rest.

He was possessed of an inquiring mind, and while admiring our cycles, questioned us about them and our experiences. What did we think of Bougie, which the French had made so beautiful? of the Gorge de Chabet, where the French had shown such engineering skill? of the ruins of Timgad, which the French had so successfully excavated? With all due regard to French enterprise, we felt like telling him that God had made Chabet-el-Akhra, and the Romans had built Timgad before the French occupation of Algeria.

Of course, the French were everywhere friendly, but did we not travel armed, as a protection against the natives? Not deeming it prudent to inform the loquacious post-master that we carried weapons of defence, we amply satisfied him by the answer, that where the French governed firearms were unnecessary; and he returned to his desk in the Bougie post-office, doubtless no less confirmed in his good opinion of the French régime in Algeria. In the universal indifference of the French colonist, we had not before noticed any manifestation of the Chauvanistic spirit. In this instance it might be pardoned in a zealous Government official.

We started for Tasmalt, trusting to luck in the matter of the inn. After a long, hilly ride we arrived there at nine in the evening. On entering the village, an Arab was employed to take us to the place about which so much fruitless inquiry had been made. The town was not large, and we soon stood before a low, one-storeyed house of small dimensions. Sounds of voices and music came from a lighted room in front.

The landlord appeared and said, –

'I am very sorry not to be able to accommodate you, but our rooms are all occupied.' Seeing our wheels, he added, 'So you are the bicyclists? Why did you not inform us you were coming, and we would have reserved a room for you?'

We replied, –

'Because we did not know of your existence. We wrote the Chef de Gare, asking if there was an inn here, but received no reply.'

'Yes, he showed me your letter, and I told him to answer you to notify us when you would come and we would be prepared for you.'

As it happened, two travellers had arrived that afternoon and taken the only two guest rooms in the house. He directed us to another house, where accommodation might be had for the night. Inwardly cursing the Chef de Gare, we followed the Arab through some dark streets to a house smaller than the first. Alas for hospitality! After we had pounded on the street door for ten minutes, a girl finally responded. She made short work of us – said everybody in the house was asleep, and slammed the door.

We returned to the inn and told the landlord we must stay there that night, even if he could only give us two chairs to sit on, for eat and rest we must after the day's journey. Seeing the necessity of doing something for us, he and his wife placed their own room at our disposal temporarily, where the latter served us with a substantial repast.

They promised to quarter us in what they called the salle à manger, but what we called the bar room, across the passage, as soon as the convivial assembly, which was making the night resonant with the sound of voices, laughter, singing, clinking of glasses, and the music of a cracked violin, should break up, which would probably be some time after midnight.

No mutton, albeit a trifle tough, and salad, ever tasted so good as that cooked by the landlady of Tasmalt. She officiated not only as cook but also as waitress, carrying a child a year old upon her shoulder the while. In the meantime, her husband was doing his best to agreeably entertain and detain his guests in the salle by constantly bringing in fresh bottles. After our meal we took each a corner of a hard sofa and essayed to doze to the ever-recurring high-pitched tune of 'Encore une bouteille.' But it was not easy to sleep, and we wandered outside every now and then to look at the sky and breathe the fresh air.

At last, about one o'clock, the revellers departed, and we went in to inspect the salle, which proved to be of the type common to similar inns in the smaller towns – a low, square room with a counter or bar on one side, over which liquors are sold, and several tables with wooden chairs. The atmosphere was reeking with the fumes of wine and smoke of bad tobacco, the peculiar odour of which combination, in spite of opened windows, lingered throughout the night with nauseating effect.

The floor was covered with ashes, stubs, burnt matches and dust. In this a bed was improvised by moving three tables together, upon which was laid one of the mattresses from the room of our entertainers. Water, and a few simple toilet articles completed the outfit, and we were left to ourselves with an apology for the absence of a door to shut off communication through a stone arch with the common passage-way. Our sleep was not of the soundest character, and we awoke at six in the morning with the impression that we had passed a night of dissipation, which impression was heightened by headache and the stale fumes that pervaded the room.

Suspecting the salle might be required for use, we immediately got up, but none too soon, for before we were fully dressed, French and Arab labourers began to drop in for early coffee before going to work. The hostess kept them in the kitchen as much as possible, but two or three found their way into the room, and stared with surprise at the two half-dressed travellers, the bicycles, the bed, and the effets de voyage scattered around. We did not blame them, for it was not every day in the year that travellers slept on tables in the public salle of the Tasmalt tavern.

The proprietor had been commissioned to have the best Arab guide of the place on hand early in the morning. While we were drinking coffee in the garden he arrived. We asked if it would be possible to get over the Col with our wheels. We knew it would be necessary to strap them on mules, because the path was rough, narrow, and ran in places along the edge of precipices, but the important point was, whether the snow at the top of the pass was sufficiently melted to permit our going over it.

He thought by this time (April 5th) it would be, although, a week earlier, two priests had had difficulty in getting through. He himself had not been up yet this season. We told him to provide the mules and we would try it. That was just the difficulty, he said. He owned thirty, but none of them were then in Tasmalt; he must see if he could find others. Promising to return in half an hour, he set out in quest of mules.

Feeling convinced that mules would be unobtainable as good lodging had been Tasmalt, in which case we should have to give up all thought of entering the Kabylie over the Col de Tirourda, we consulted our map for the best route around the Djurjura mountains.

The Arab returned with the report that mules were to be had, so bidding farewell to our host and hostess, who had done all in their power to make us comfortable, we set out to skirt the mountain wall which separated us from the Kabylie, determined to carry out our project, even if it took three days to reach the entrance to the region.

Notwithstanding the comparatively sleepless night, we made that day, under threatening skies and in some rain, one hundred and twenty-four kilometres, and reached Menervill at six in the evening. During this day's ride we overtook an Arab cavalier on the road – one of the better class, who are usually seen mounted on short, thick-set, shaggy horses of rather dumpish appearance, and armed with long, old-fashioned guns, such as are carried in Eastern countries.

THE CAVALIER.

As we passed this horseman he seemed inclined to try the mettle of his horse against our wheels. The grade at this point was down and the road excellent, so, as we had never witnessed the pace of a horse such as he rode, we accepted the challenge. As he increased his speed we increased ours, till a velocity was attained that no horseman could safely exceed on a descending grade, while we were still within the limit of our speed. Letting our machines out a little more, we left him behind.

A few minutes later, when the level was reached, the clatter of hoofs was heard behind, and we saw he was going to make another trial. We quickened our pace again, but this time the physical advantages were on his side, and in a twinkling he flew by us like a whirlwind, brandishing his gun in the air, over his head, with his right hand. We shouted 'Bien fait' as he passed, which seemed to please him. After two or three hundred feet he drew in his horse and resumed the customary slow trot, saying, with a smile, something in Arabic, which we did not understand, as we rode on ahead. Our respect for that breed of horses was greatly increased by this exhibition of its capacity.

The next day we reached Tizi Ouzou, fifty kilometres, by eleven o'clock. This is the first important military station in the Kabylie, and is prettily situated among rolling green hills. From this town, with the exception of a few kilometres of level, the rise is continuous to Fort National, which lies over six hundred metres higher.

The trip to Fort National, whether taken on foot, in a carriage, or by diligence, is essentially a walking trip, and so it happened we fell in with a French officer who was, ostensibly, going to the Fort by the post, but practically, as we were, afoot. Were Monsieur and Madame going up to visit the Fort? No, we answered, we were going to see the Kabyles. They were a bad race, he said, but the country was worth a visit. Did we know the brigand chief Areski was in prison at Tizi Ouzou? Yes, we did, and we also knew he had been held there for a considerable time, the courts, with their customary indecision, being unable to determine what to do with him.

As we walk towards Fort National, some facts about this famous brigand may not be out of place.

Areski and his band have been for years the terror of the forest region between Fort National and Bougie, and, indeed, of the whole Kabylie, where, in most of the larger towns, he had houses, on which he paid the Government taxes regularly.

Roving with his band over the country, wherever he saw a house, a farm, cattle or other objects he desired, he would declare himself the owner and take possession as such, giving the real owner to understand that no claims were to be preferred against him, under penalty of death, which was certain to follow if his orders were disobeyed. He placed agents of his own over the farms so claimed, who managed them in his interest, accounting to him for all the profits. If they proved false to him, or in any way unfaithful to their charge, a shot from his revolver ended his business relations with them.

He exacted tribute freely, but with a certain regard to the means of his victims, and it is said he seldom demanded more than they could pay. Towards the poor he showed generosity, relieving their distresses and aiding them with timely gifts. Although on occasion he plundered and burned and devastated, yet, as a whole, he displayed a knightly tact in his exactions, which made the natives not only stand in awe of him, but also, respect, and sometimes even, love him.

One of Areski's most engaging qualities was his chivalrous hospitality to strangers, whom he received in the forest with friendliness, feasting them on game he had killed himself. He and his band prided themselves on their code of honour, one article of which was never to molest strangers. His age is a little over forty years.

The head of a second band of brigands of much less fame, named Abdoun, committed a murder in the execution of blood revenge, which custom exists among the Kabyles as in Corsica and Sardinia. For this he was transported for life to Cayenne. His family sent him what money they could procure, and, awaiting his time, he escaped to England and thence to Morocco, from whence a journey of six hundred kilometres brought him, after five years' absence, to his native land.

He placed himself at the head of a robber band, and united with Areski and another chief for the sole purpose of attacking the tribe of Tabaruss, against which he had sworn vengeance. This attack, which resulted in the breaking up and annihilation of nearly the whole tribe, at last induced the Government to really exert itself to put down lawlessness.

After several attempts and considerable bloodshed, the efforts of the seven hundred French soldiers detailed for this service were rewarded by the capture of Areski and forty-two of his band. The rest fled and discreetly disappeared. This happened over a year ago, and still the criminals remain untried, a fact which has not raised the French Government in the estimation of the more intelligent among the Kabyles.

Thus Algeria, although it cannot perhaps, like Italy, claim to be the classic land of banditti, may be called the home of clever Fra Diavolos, who, up to the present time, under the eye of the Government have pursued their calling in the most approved manner.

Areski has a pretty wife, called Tessadit, with whom he is said to have lived amicably, though neither party was particularly true to the marriage vows. She lived with her father-in-law at Bou Hinni, where she owned a fine house. The Kabyles continued to pay tribute to her after Areski's capture, as they had done previously under his orders.

But fair Tessadit was avaricious, and having found another husband, she no longer cared to support her father-in-law. One day, when the old man was returning from the field with his mule, he received two bullets in his head, and so did not return home alive2.

All this time the road has been leading us higher, winding in great curves around the mountain, and lastly over a long crête, at the top of which appear the gates of Fort National.

CHAPTER XV
THE KABYLIE OF THE DJURJURA – PECULIARITIES OF ITS INHABITANTS – HISTORY

LOOKING across the bay to the east, on a February afternoon, from a balcony of one of the luxurious hotels of the Mustapha at Algiers, the eye is arrested by the picturesque mountain range of the Djurjura. Draped in a burnous of winter, it rises proud and beautiful to where its pointed summits are lost in the roseate clouds of evening. At Fort National, one hundred kilometres nearer, the picture changes, and a scene unique in the world opens up before one. Across an intervening space of twenty kilometres rise the same peaks, with weird, verdureless, sharp outlines capped with snow. Running forty kilometres in length, they reach their greatest height in the Lalla Kredidja – twenty-three hundred and eight metres.

This range, called Mons Ferratus by the Romans, and regarded by them as impenetrable, guards, like a gigantic fortress, the multitudinous crêtes of the Kabylie of the Djurjura. Unlike many abrupt peaks which, after the first talus, descend in commonplace meadow or forest slopes, here where the wall ends the spur begins. These spurs or crêtes, running in wild irregularity, fill the space between the Djurjura and Fort National, and upon their narrow crests the Kabylie villages are built, four or five, or even more, upon a single height. On some of the mountains, at their junction with the crêtes, cedar forests are found, but never lower than a thousand metres.

Gardens filled with olive, fig, cork, oak and ash trees, walled in by prickly pear and laurel, follow at a short distance upon the barren rocks and forest trees of the higher regions. All this semi-tropical profusion is kept green by Mons Ferratus, which in its spring bounty sends forth scores of laughing rivulets upon the Kabylie.

The spurs shoot out in such numbers and with such persistence, that wide valleys are unknown. The villages are reached by steep winding paths overhung with most luxuriant vegetation. In its grandeur, its garden-like beauty and originality, this region can be compared with no other. Taken singly, the Djurjura might be likened to the Alps or Pyrenees, but let the eye fall lower, and what can be compared with the crêtes of the Kabylie?

The climate is like that of other interior high parts of Algeria, more rigorous than that of the coast. The villages and towns vary in altitude from eight to twelve hundred metres, and accordingly, a freezing temperature is often experienced in winter. Snow sometimes falls, though rain is most common to that season. The most delightful months, when the world below Djurjura is like a garden of Paradise, are April, May and June. By June, the noonday heat becomes oppressive, but the nights are cool, as they are said to be throughout the hot weather.

If their country is curious and unique, the Kabyles of the Djurjura, direct descendants of the Berbers, are no less interesting. As to whether these Berbers, one branch of whom conquered Spain, came from the Aborigines, who were of mixed races, or were lineal descendants of the Numidians, authorities differ. To those who have studied the variety of types in the Kabylie, where blue eyes and red hair, black eyes and swarthy skin, brown eyes and bronze red faces with almost Ethiopian features are seen, the former supposition seems most plausible.

These hardy mountaineers of the Djurjura, unlike the Kabyles of Morocco or southern Algeria, seem to have preserved the warrior spirit of the Berbers, and in their mountain fastnesses remained unconquered by Roman, Moor or Turk. They first appear in history as a race under the Romans, who invaded their country but gained no foothold. The Arabs, under the Khalifs, established themselves temporarily there. The influence of their presence is seen in the adoption, by the Kabyles, of the religion of Mohammed, so far as they adopted any creed, and in the language, which is a mixture of Arabic and old Lybian.

The term Kabyle, applied by the people to themselves, is derived from the Arabic word Qebaiel, meaning league or clique. This name symbolises one of the most marked characteristics of the people, who from early times have been divided into cliques or tribes, to which the first allegiance of the individual is given. The existence of a strongly clannish spirit, more than is usual in other tribes in North Africa, led to discord and constant warfare among themselves up to the time of the French occupation. Although the religion and customs are practically the same throughout the Kabyle race, yet they like to speak of these as having rather a tribal than a racial significance. For example, a member of the tribe of Beni Yenni, which to-day occupies three villages, being asked as to certain Kabyle customs, replied 'With us, at Beni Yenni, we do so and so.'

As against outside interference with their affairs, however, they were united and of one mind. M. Aucapitaine, speaking of the relations between the Arabs and Kabyles, says, 'Ils n'ont qu'on point de contact, leur haine réciproque.' Similarly, in considering the relations of the Kabyle tribes to one another, it might be said they had one trait in common – the desire to exclude all other people from their country. Their motto would appear to have been, a land divided against itself, but inhabited by their own race, rather than a land controlled by Romans or Arabs.

Another well-marked trait has been their devotion to self-interest, which has generally appeared as a leading motive in their dealings with outside people. They allied themselves with the Roman, the Arab or the Christian as the case might be, the only consideration being that advantage should accrue to the Kabyle as the result of such alliance, and they lent their arms to that side which they thought would reward them most liberally.

The French claim the honour of being the first to conquer the Kabyles, but the victory was not won without severe fighting and great loss of life. Shortly after landing in Algeria, they besieged and, with little difficulty, captured Bougie, but from the moment the garrison was established there, the Kabyles made their lives miserable, and depleted their numbers by frequent attacks. The French finally gained the upper hand, but only after losing one of their generals, Salomon, who was treacherously murdered at a rendezvous with a sheik, just outside the city.

After hard fighting, the French, under MacMahon, gained possession of the Grande Kabylie in 1857. Unable longer to hold out against the improved methods of warfare and superior numbers of their opponents, the sheiks of the different tribes visited MacMahon in his tent and submitted to the authority of the French, and agreed to pay an indemnity of one hundred and fifty francs a gun.

The French at once began the erection of Fort Napoleon, now called Fort National, at Souk-el-Arba, on the long crête facing the Djurjura and commanding the villages of a number of tribes. The timely building of this fort made a strong impression on the people. While the foundations were being laid, an old Amin of one of the neighbouring villages came to pay the tax of the community, and seeing for the first time the foundations of the fort, addressed the following remarks to an officer present –

'Is the Sidi Maréchal going to live at Souk-el-Arba?'

'No, he is having a bordj (fort) made.'

'A bordj? Ah, then they told me the truth. Look at me. When a man is about to die, he returns to his house and closes his eyes. Amin of the Kabyles, I close my eyes, for the Kabylie is going to die;' and for several minutes the old Amin remained standing with closed eyes.

This truly Oriental figure of speech expressed sadly and simply the fading power of the Berber race.

The building of Fort Napoleon continued, and the great turnpike connecting it with the plain and Tizi Ouzou was made. MacMahon remained some time longer till all the tribes, one after another, were conquered. It took more than one campaign to transform all the Kabyles into Frenchmen. Finally, one of the most war-like and remote tribes, the Zouaouas, yielded, and last of all the mountain towns of Takleh and Tirourda, over which presided the venerated prophetess, Lalla Fathma. From the Zouaouas, who afterwards ceased to exist as a tribe, comes the name zouave.

When his work was finished, MacMahon, with a small company, traversed in safety the whole Djurjura region on his return to Algiers. About this time a French soldier accosted an old man of the Illilten tribe, saying, 'You are a friend at present, old man.' 'Friend – yes, friend because I have no powder.'

Afterwards, when the powder was again procured, the French had to take up arms anew against a land they thought they had conquered. In 1864 and 1866 slight revolts, easily quelled, occurred. In 1871 an insurrection of greater magnitude took place, for the suppression of which troops fresh from the Franco-Prussian war were called into requisition. At the time of the German war Algeria was tranquil, and both Arabs and Kabyles furnished troops which fought bravely, ten thousand of them falling at the battles of Reichshofen, Sedan and Orleans.

Several causes led to the insurrection of 1871, for which the French themselves appear to have been to blame. In 1870 the Kaids and other chiefs of the land offered their services and means to Napoleon, the rejection of which offer rankled in their breasts. Then came the naturalisation and elevation to office of the Jews, which added to the discontent. Much as the Kabyle dislikes the Arab, he dislikes the Jew more, and when he saw him raised to political power, and acting on the jury, his wrath could no longer be restrained. The radical press also attacked the Arabs and Kabyles in an unwarrantable manner.

This insurrection lasted over three months, and was attended with great loss of life. For certain treacherous dealings, a number of chiefs were publicly executed at Algiers, and the Kabyles alone were obliged to pay two hundred and ten millions of francs. This experience showed them that it was useless to contend with a power that was too strong for them, and for ten years quiet reigned under the walls of the Djurjura.

In 1881 a religious fanatic tried to induce the Kabyles to rise, but with little success. They remained mostly indifferent, unwilling to incur anew the disastrous experiences of 1871. The Kabyle had become convinced his best interests did not lie in making war against the French.

CHAPTER XVI
KABYLE VILLAGES-CUSTOMS – INDUSTRIES – EDUCATION

THE villages are long and narrow, running along the tops of the crêtes, and extending but a short distance downward. They command a view in all directions. The houses are small, of one storey, made of stone and roofed with tiles. Those on the outside of the town are put up one against another, with no openings towards the outside, thus forming a wall of defence. There is but one street of entrance, and across this, at short intervals, are thrown stone archways, which in former days were provided with heavy doors. At the centre of the villages some of the dwellings have small gardens around them, hedged with prickly pear.

The houses consist simply of four walls and the roof, are usually without windows, and besides the door have an opening in the roof. Beneath this is a hollow in the earth lined with stones, which forms the fireplace, the sacred spot of the house, where strangers and friends are received and entertained, cheered by the blaze of roots and branches.

In one corner the earth is scooped out so as to form a depression below the level of the rest of the floor. This serves as a stall for young calves or donkeys. In another corner a home-made cradle, consisting of an old piece of burlap strung on two poles, holds the last born of the household, watched over by its grandmother, who is seared and wrinkled enough to be its Berber ancestress.

Furniture a Kabyle interior does not possess. A few plates, and iron and clay pans and kettles for making couscous, complete the kitchen outfit. Such is the Kabyle house to-day, and such it has doubtless been since the time of the Romans.

Before the French invasion, the Amin was the chief man of each town, and all disputes were brought before him at the Djemaat, where the town council was held. This was a rectangular structure, provided within with stone benches. To-day, the head men, or mayors, are French, and Kabyles are employed under them.

The judicial department of the Kabylie, as of Algeria in general, is in a far from satisfactory condition, as is illustrated by the following incident. A native bought a piece of land, for which he paid four hundred and thirty francs. The ownership, as often happens elsewhere, was disputed. In the course of legal proceedings he went to Constantine ten times, travelling each time two hundred kilometres. After the tenth trip, besides having travelled two thousand kilometres and spent six hundred francs, he was still uncertain whether his title to the lot was valid. When Frenchmen discuss these matters they smile, shrug, and remark, – 'Que voulez vous? C'est le même partout.'

Among the occupations, agriculture holds a high place. The plough is a sacred object, and the manufacturer of it practises a holy calling. It is a very primitive and inefficient implement, similar to that used by the Arabs. The Arabs are poor farmers, and in ploughing scarcely do more than scratch the surface of the ground. If a stone of any size lies in their way, they plough around it rather than remove it. Their harvests are correspondingly small. The Kabyles are much more painstaking, pursue agriculture with more method, and reap a greater reward.

Barley is the chief cereal, figs and olives the staple fruits. Figs form an important item in the diet. Olive oil is used in a variety of ways in cooking, and in the household economy. Sorghum is also largely raised. The cattle are fed almost entirely on the dried leaves of the ash tree, gathered in August and housed for winter use in little open houses constructed of poles, with conical thatched roofs.

One of the chief industries is the making of olive oil, which might become a source of much profit were not their methods so defective. Only a native palate can endure the article produced. Many of the people, recognising the imperfection of their own methods, sell their olives to the French oil manufacturers.

The women, besides doing all the housework, make the pottery – the jugs, jars and water-pots, and the graceful amphorae. In the manufacture of tiles, both men and women engage. To women's work belongs also the weaving of linen for shirts, and of woollen cloth for the burnous.

A very important industry is the manufacture of silver and metal ornaments, in which the tribe of Beni Yenni is especially skilled. They make an interesting variety of pins, bracelets and boxes, which, when made from the native designs, are curious, and often highly ornamental. In bracelets, patterns from the Paris boulevards have of late crept into vogue among these mountaineers, who fail to appreciate the fact that the Kabyle designs are more original and attractive than imitations of Parisian trinkets possibly could be. The pins, bonbonniéres and spoons, made of silver and enamel, and studded with coral and turquoise, and the large silver drinking cups with handles, are most sought for.

The ornaments commonly worn by the women are made of brass, enamelled and inlaid. Those for the market, of the same designs, are of silver. The richer women also wear silver ornaments, especially on fête days. The enamel is made of coloured glass pulverised and mixed with water. The design is drawn by soldering on a silver or metal thread. The liquid enamel is then applied, and exposed to sufficient heat to cause a proper fusion.

Before the French conquest the Kabyles resisted all taxation, paying only a certain sum to the village treasury, and for the relief of the poor. Since 1871 taxes have been levied in the following manner – Indigents pay nothing, those who have a little property, ten francs, those in easy circumstances, fifteen francs, the rich, fifty, and the very rich, a hundred francs.

They are a hospitable race, and pride themselves on treating strangers well. If a traveller passing through a Kabyle village has no introduction, he has but to say to the first man be meets, 'I am here as guest of the town,' and he is at once taken to the mayor, who sees to his proper entertainment. In this land, where there are no inns, the guest is regarded with especial deference, and has nothing to fear, for he is under the protection or 'anaia' of his host, and of the whole village, which extends to him its hospitality.

If a Kabyle accords his 'anaia' to a person, he is obliged to accompany him, if he possibly can, to the end of his journey. If this is impossible, he must give him a gun, a dog, or some object which can serve as a safeguard on his journey. If the 'anaia' is violated during the journey, the one under whose protection the guest travels, his kindred, or members of his tribe, are under obligation to avenge the insult.

The 'anaia' and the vendetta, which is still practised in the Kabylie, go hand-in-hand, and it is difficult to say which has been the cause of more combats – the violation of the laws of the 'anaia' or the pursuit of the vendetta. Daumas says a Kabyle will abandon his wife, his children and his house, but never his 'anaia.' This custom, unlike the vendetta, is said to be practised only in the Kabylie.

The Kabyle is as punctilious in matters of charity as in his 'anaia.' The famine of the winter 1867-8 furnished a memorable instance of the display of this admirable characteristic. Starving people from all over Algeria, and even from Morocco, sought and obtained succour in the Kabylie. They were fed and clothed without regard to nationality. This was done, as are all their charities, without ostentation, but so thoroughly, that it is said not one person died of starvation that winter in the Kabylie.

It is regarded as a sacred duty to provide for the poor, who have the right whether native or stranger, to sleep in the mosque or the djemaat, and, if hungry, to stop at meal-times at whatever house they please, where they will not be refused a plate of couscous.

In educational matters the Kabyle appears unable to go beyond a certain point, hence little education is attempted by the French except in the elementary schools. Up to twelve or thirteen years he is quick to learn, and often shows better memory than the European boy of his age, but after that his facility is less and his faculties become dormant. He is unable to reason or comprehend the abstract.

The boy is very glad to leave school at fourteen, after which, surrounded only by native influences, he soon drops any French habits he may have acquired, and with them the French language and its attendant accomplishments of arithmetic, geography, and the Fables of Fontaine. The last is memorised, as are verses from the Koran, and with about the same result. But it is not necessary to look among half-savage races to find little girls and boys who have found it hard work to understand La Fontaine.

What a generation or two under French tutelage will develop, time will show, but just now the influence of the Numidians and Moors is more apparent than that of the French. Awaiting, however, the day when education, the only real civiliser, shall make a deeper impression on Kabyle character, the French take advantage of the decided bent of these natives for business. They are naturally sober and industrious, two excellent characteristics which make them valuable subjects, but which are so lamentably wanting in their Arab neighbours.

The Kabyle has been called a fatalist, but this is scarcely true. 'Mektoubrebbi' he cries when a great grief comes upon him, which has been translated – 'It was written,' but which more correctly is said to mean – 'God has willed it,' and, thus translated, it loses its fatalistic meaning. Having met with misfortune, he is philosophical and does not waste his time in vain regrets, but seeks to make the best of what is left.

CHAPTER XVII
FEMMES KABYLES

As the advance of a nation in modern ideas may be judged by the position occupied by its women, Germany being a notable exception, so we may form some conception of the degree of civilisation existing in the Kabylie at the present time, by a brief consideration of some of the customs pertaining to the women.

Writers, guide-books and the French tell us that the women of the Kabylie occupy a high position, and one much more favoured than that of Arab women; but those who have seen them in their homes do not all agree with this opinion. As a proof of her emancipated position, it is urged that the Kabyle usually has only one wife. The fact is true, but it may be accounted for on economic grounds rather than as being due to moral principle. Nothing in the laws or religion forbids his having as many wives as he can support, and those who do business in Algiers in winter, and return for agricultural pursuits in summer, generally follow Oriental custom and have a wife in each place. Again, it is said the Kabyle woman goes with uncovered face. This is simply in accordance with long established custom, for from time immemorial only the wives of marabouts have covered their faces. One must look deeper than this and see how man regards woman, and how woman regards herself in Kabylie land.

The opinion the Kabyle holds of woman may be seen in the following remarks made by a native to a French captain: –

'When two children are born, you would surely have me show more joy at the birth of the one that will be able, gun in hand, to defend his tribe, than at that of the other, who at the best will only be fit to make bad couscous?'

'But,' replied the captain, 'these daughters you so despise are in reality bargains, since, at the age of ten or twelve years, you sell them for a good price.'

'Yes, but they do not give me the influence in the council that being the father of four or five well-grown sons does. Besides, our laws make a difference between the adult male and female – for example, a wife cannot kill her husband, whereas a husband may kill his wife if he deems it necessary.'

The birth of a son is a cause of great rejoicing, and is announced by loud outcries from the women of the village, to which the men respond by discharging guns. After this proceeding, jewels, perfumes, cosmetics and stuffs are brought as presents to the mother. A few days later a fête is given. The mother, with great pride, decorates her brow with the 'thabezinth,' a circular silver or metal brooch, about four inches in diameter, set with coral, which she wears a year and then removes, placing it again on her forehead only in the event of another son being born.

On the contrary, the newly-born girl opens her eyes on a world opposed to her advent. Her birth is celebrated in a very quiet manner, and only by the family circle – as the Arab would say, without noise. No presents are given. Women having daughters may wear the thabezinth also, but only as a breast ornament.

As the girl grows out of infancy, her education amounts to little. She does sometimes attend the village school, but exceptionally. Kabyle parents are opposed to having their daughters educated à la française – that is in school – for the reason that it lessens their chances of marriage, no Kabyle desiring a wife who has been to school, however little she may have learned. Not long since, a chief lamented bitterly his folly in allowing his daughters to attend school, as he had not been able to procure husbands for them, and they, at the age of twenty-five, were still on his hands. Their chances of matrimony were indeed small at that age – vieilles filles of fifteen years' standing.

At twelve years, an age when girls in civilised countries begin to study, the Kabyle girl, without even rudimentary knowledge, with all the innocence and ignorance of childhood, is sold to her husband by her father, who considers himself very fortunate if her charms bring him five hundred francs, the average price being two hundred and fifty.

A friend of the bridegroom makes the arrangements with the bride's father, and when all is satisfactorily settled, the future husband and his friends run about the village firing guns to announce the happy news. This is the first the girl knows of the matter, about which she is in no way consulted. A second interview follows, in which the lover pays a part or the whole of the purchase-money through an agent.

Marrying a wife is expensive in this land. Besides paying for her, the husband is expected to present her with a panier de la fiancée, consisting of seventy different essences, medicines and cosmetics. At the preliminary meetings neither bride nor bridegroom are present.

Nor are they present at the marriage ceremony. Before pronouncing the absent couple man and wife, the marabout, in the presence of the family and friends, reads the following enlightened words from the Koran: – 'Men are superior to women, because the qualities which God has given them elevate them above women, and because they use their money for the marriage dowry of women. Virtuous women are obedient and subdued.' The advice is given to a man about to enter the matrimonial state – to reprimand his wife if he fears disobedience on her part, to beat her if she disobeys, but from the moment that she obeys, to cease quarrelling with her.

After the ceremony the bride, for the first time, appears on the scene, dressed in her richest garments, and closely veiled. She mounts a mule, and is conducted to her husband's house. At the door, where he awaits her, a vase of water is handed her, with which she sprinkles the assembled people. This is a very ancient custom, the significance of which is now unknown. A basket containing couscous, nuts and cakes is next handed her while sitting upon the mule, and these she throws to the crowd. In the basket is an ankle bracelet, which she keeps, this being a symbol of the chain of wedlock she has assumed.

At her entrance on married life, the Kabyle woman is often a pretty creature, but under the weight of domestic cares she soon grows old. Her duties, besides childbearing, comprise all the housework, the weaving of cloth, the making of couscous and, hardest of all, the daily bringing up of the water for household use, from springs on the mountain side or in the valley below the village.

At certain hours every day lines of women may be seen, with heavy earthen water-jars on their backs, toiling slowly up the narrow, almost perpendicular, paths to their houses, clad in a large white tunic, fastened on each shoulder with heavy silver pins, and held at the waist by a woollen belt, with legs and feet bare, except for three or four ankle bracelets, their long arms thrown backward supporting the amphora. They are eminently classic in appearance, though barbarous in their thraldom.

For this life of continued drudgery she is rewarded by being permitted to eat in the presence of her husband, and when time and care have rendered her no longer attractive to her lord, she is replaced by a younger, fairer rival, who kindly accepts her services as servant.

The condition of the Kabyle women to-day, but slightly better than that of the Mauresques, is about what it was a hundred years ago, and what it may perhaps be a hundred years hence, for they are utterly without hope, and have themselves no power to better their condition. The idea that they have a better position than Arab women doubtless comes from a misapprehension of existing circumstances.

It is believed, by those who have studied the matter, that in the time of the Berbers women held a higher place than now, and that their position became debased after the adoption of Islamism.

The Koran permits the Kabyle, as well as others, to have as many wives as he chooses, but he rarely avails himself of this permission, at least during the early years of married life, for the reason that it costs much to buy a wife and more to support her.

CHAPTER XVIII

A DAY IN THE AFRICAN ALPS – THE COL DE TIROURDA

THE town of Fort National stands on the crest of a long narrow crête running parallel with the Djurjura, with a commanding view of the country on both sides. The town is mostly French, and presents nothing of especial interest except the Wednesday market. The fort, on a hill overhanging the town, is quite an imposing structure, suitable to the importance of the place as a military post. It was enlarged after the insurrection of 1871. At that time it was termed by the natives, 'A thorn in the eye of the Kabylie,' and such they probably still consider it, as it is kept well garrisoned, and, in case of insurrection, the houses of sixty thousand Kabyles could be destroyed in a few hours by its guns.

While looking about on the afternoon after our arrival, we chanced to go into the shop of the only bijoutier in the town, and while examining Kabyle ornaments we remarked, in the course of conversation, we should like to visit the place where they were made. The jeweller, a very friendly Kabyle, who spoke French, said it would give him much pleasure to entertain us as his guests at his house at Beni Yenni, if we would come there. He added, –

'To-morrow is the beginning of the feast of the Ramadan. I am going home early in the morning to remain with my people four days. Will you not go with me?'

'No,' we said, 'we cannot go to-morrow, for, fair, we have arranged a trip to the Col de Tirourda. Perhaps the day after to-morrow.'

'As you please. I will be there.'

'But how far is Beni Yenni from here?'

'Four hours on a good mule.'

As we looked at the fine form before us, wrapped in the soft folds of his burnous, which fell to his sandled feet, his face bright, his eye eager with apparently real desire to have us visit him, a strong wish to see the land of the Beni Yenni seized us, and we promised to go on the second day after.

That night it rained, as it can in the mountains, in torrents, and we feared our attempt on the Col de Tirourda from this side would prove as unsuccessful as the one proposed from Tasmalt. But at five o'clock in the morning, when a small Arab boy knocked at the door to awaken us, the brilliant African stars were palpitating in the heavens – for stars do not twinkle in Algeria – and the dark blue sky was full of mysterious premonition of a cloudless dawn.

We decided to use our cycles as far as the road would permit, and then climb the Col on foot. No one at Fort National could tell us anything about the route beyond Michelet, twenty-five kilometres distant. As we rode out of the sleeping town, past the house of the bijoutier, a cloaked figure appeared at the door, and the words 'Je vous attendrais' greeted us. We answered, 'A demain a Beni Yenni.'

The road, though well made, was muddy after the rainy night, and we had to ride with care, but we would not have hurried if we could on this magnificent ride from Fort National to Michelet. The road runs along a high horizontal crête, and as it circles in and out of the curling contour of the mountain, brings into sight peak after peak of the Djurjura, and ridge after ridge of the village-covered Kabylie. We could not help recalling the road from Sorrento to Positano and Amalfi, and that from Taormina toward Messina, not on account of any similarity in the landscape, but because each in its way is so superlatively lovely.

We passed a number of villages, and many Kabyles on the highway returning home from the market of the previous day. Most of them carried pieces of meat strung on strings, for the long fast was over and the people were about to celebrate the feast which was to follow by free indulgence in meat and couscous. Some of the men were without the usual cloak, being dressed only in large yellowish white shirts confined at the waist by a belt, their legs and arms bare. The only addition to this costume is the cloak ordinarily worn in cold weather, or sometimes two or three when the cold is extreme.

Michelet is a small official station consisting of a few houses and a sort of fort or strong house to which the inhabitants may fly for protection in case of necessity. Its situation, facing the highest peaks of the Djurjura, is even more beautiful than that of Fort National. Fifty villages are seen from here. On one long spur running out from the mountain we counted fifteen, the first nestling on the limit of the green, overhung by the high crags above, the last perched on the end, where it fell off sharply to the valley below.

The knowledge of the inhabitants of Michelet as to the road did not extend beyond the next station, the 'Maison Cantonniere,' nine kilometres farther, to which point they thought we could ride. They did not tell us that the road was unfinished, and converted by the rain of the preceding night into a sea of mud, as we found to be the case after going on a short distance. Returning to Michelet, we left the cycles at the inn and started on foot for the 'Maison Cantonniere,' which, by picking our way along the sides of the road, we reached in two hours.

Here we found only two tattooed, bejewelled Kabyle women, who spoke no French, and could tell us nothing about the route to the pass. We were now well up on the mountain side. The road, or the attempt at one, ceased, and we struck into a mule path, which led directly upward around the edge of the mountain. The scenery had become Alpine in character, the fertile lands were left behind, and the region grew milder and more desolate with every step. The path was obstructed in places with débris and rocks, which had been loosened by the frost and rains and had rolled down from above, so that progress was oftentimes slow. In one place we had to climb over a chaos of good-sized boulders, which covered the path for two hundred feet, the result of a landslide. This was the route across the mountains which we had been told in Algiers was available to bicyclists. We found it difficult enough to follow on foot and were thankful our machines were safely housed.

A turn of the path now brought into view the village of Tirourda far below, on the top of a desolate crête, the last village of the Kabylie in this direction, and the last to fall into the hands of the French in 1857, noted for having been the home of the Berber prophetess, Lalla Fathma.

Built almost upon the snow-beds of the mountain, amid nature's sternest aspects, it looked more like a deserted eagle's nest than the home of man. In 1857 it and Takleh, an adjoining village belonging to the marabouts of the tribe of Illilten, were governed by Sidi Thaieb, who, with his sister Lalla Fathma, lived like a petty monarch, surrounded by every luxury. Lalla had a wide reputation as a prophetess, and was consulted by the people of the whole Kabylie.

At the approach of the French, Sidi Thaieb, whose family had for centuries governed these two villages, went to MacMahon and succeeded in making an arrangement by which, in return for certain services on his part, these villages should remain unmolested. This compact would have been carried out by the French had not an unfortunate mistake occurred. Several zouaves in pursuit of some fugitives from another town came, without realising the fact, into Tirourda. The people, supposing them to come with hostile intent, fixed upon them, killing and wounding them all. A detachment of the army, which was in the neighbourhood, hearing the firing came up, and before either side was aware of it, a general engagement took place.

Lalla Fathma and her brother were taken prisoners, and their arms, clothing, furniture and jewels were confiscated. As we sit on a boulder, looking down upon lifeless Tirourda, fancy pictures the procession of three hundred men and women, with torn garments and dishevelled hair, starting down the narrow steep path on their way to the marshal's camp, following their idol seated on her mule, who, in the midst of the tumult of battle, had found time to braid carefully her jet black tresses, paint her cheeks with carmine, blacken her eyelids and stain her finger-nails with henna.

A light rustling breaks the silence. Is it the voices of the sobbing women and children driven from their homes by the relentless conqueror, or is it the spring wind playing about the jutting crags of the Pic de Tirourda?

The path became narrower and ran along dangerous precipices. Evidences of the action of frost appeared in cracks on its outer side, where the loosened edge threatened to slide off in places should any extra weight be placed upon it.

All this part of the mountain is composed of a loose, crumbling material, which at this season affords an uncertain and precarious foothold. Patches of snow lay across the path, and were seen to cover it ahead in great slanting sheets running to the edge of the cliffs. These must be crossed if we would go on.

We had now arrived within two kilometres of the goal in regard to which, we had so long sought for information in vain – the Col de Tirourda – from which an extended view over the Kabylie of the country south of Djurjura of the sea, and of Algiers was to be obtained. Should we go on, or give up farther attempt? It was hard to give up now, after so much effort and being so near to success, but common prudence dictated that we go no further. We knew nothing of the character of the ground under the slanting masses of snow ahead, or what foothold they had upon the mountain. We had no Alpine-stocks, and no hobnails in our boots. Reluctantly we turned our backs upon the Col, feeling that we could at least give somewhat more information about the route than we had ever succeeded in gathering from others. The grandeur of the region, however, amply repaid all exertion, and we consoled ourselves for the failure to accomplish all we had desired, by the reflection that we were not the only tourists who had had mountain luck.

We were now thankful that the mules had failed to put in an appearance that morning at Tasmalt, for had it been otherwise, we might have been deposited with our bicycles on the top of the Col, there to wait perhaps until the snow melted.

By the time we reached Michelet again we had walked eighteen miles up and down, and were glad enough to mount and ride back over the splendid road to Fort National. Over the green setting of dozens of villages hovered in the evening light a thin blue film of smoke, a sign that the labours of the day were over and the people had gathered around their hearthstones to enjoy their evening meal. The clouds behind the mountains grew pink in sunset hues, and faded into the cold greys of crepuscule before the day's trip ended.

CHAPTER XIX
AMONG THE KABYLES OF BENI YENNI

THE next morning at six two mules, accompanied by two Kabyle guides, stood in front of the Hotel des Tourists ready to take us to Beni Yenni. Women's saddles appeared to be unknown in that region, so we accepted those furnished and both rode astride, which is certainly the safest way on the steep rough paths which have to be travelled. The saddles having no stirrups, we suggested to the guides that our comfort would be promoted if some contrivance could be arranged by which we could brace our feet. Accordingly, a grass rope with a loop at each end was thrown over the saddle. This answered the purpose fairly well, but some dexterity was required to preserve a proper equipoise of the feet.

Soon after leaving the town, the path began to descend in sharp zigzags along the side of the Souk-el-Arba spur. To us who were unaccustomed to this manner of exercise, riding on one of these long-legged animals down a steep grade was not wholly agreeable. Its shoulder-blades moved up and down to a marvellous extent, causing a jolting that would have done credit to any of the apparatus of a movement cure. Its body was pitched to such an incline that constant attention to the rope stirrups was necessary to prevent sliding off over its head. Added to the unpleasant sensation of plunging downward, the mule persisted, particularly at the sharpest curves, in walking on the very edge of the path. The attention of the guides was called to this fact, but from their small stock of French was gleaned only the comforting information, 'Il va bien, sur très sur.'

Considering the mule was accustomed to doing just that sort of work every day of its life, we concluded to let it go as it pleased, and direct our efforts to holding on.

We passed through two villages built on lower projections of the crête. As we descended, the vegetation became more luxuriant and the path more picturesque, hedged by laurel, hazel and blackberry bushes, and lightly shaded by the young leaves of the ash, almond and chêne liège. In June the country must be a bower of green. Every few turns the path crossed the rocky beds of cascades bounding towards the deep valley below.

At some of the brooks women were washing clothes in the running water, treading upon them with their feet, as is customary among the Kabyles as well as among the Arabs, the music of the bangles upon their ankles mingling with the babble of the water. This method of washing does not offend the taste more than that employed in many parts of Europe, where the articles washed are rubbed against stones in canals and streams; but when it comes to preparing salad in the same manner, as is also done in the Kabylie, the impression produced on the minds is unsavoury.

Men on mules met us. How infinitely picturesque a burnoused Oriental can look on a mule, and how amazingly commonplace a European. Lines of women, bearing amphorae filled with water, filed slowly past, and sometimes a mother with two or three pretty children would stand aside for us. With nature so beautiful and people so interesting, it was like a pastoral symphony composed by a musician of ancient days.

We looked for our Kodak to photograph some of the interesting groups, and to our dismay found it had been left behind in the hurry of departure, which accident we have never ceased to regret, as the best opportunity of the whole journey of photographing characteristic and peculiar types and costumes was thus lost.

On reaching the bottom of the narrow, blooming valley, a stream was crossed and the path began immediately to ascend another crête. The guides gave us to understand that climbing was less disagreeable than descending, which proved to be the fact; but before the four hours were over, our ambitions in regard to mule-riding were satisfied, and we were relieved to see at last the walls of Ait l'Hassen, the largest of the towns occupied by the Beni Yenni, gracing the height just above.

Ait l'Hassen has about eight thousand inhabitants. Our mules clattered through the narrow streets, followed by many of the inhabitants, and soon brought us to the house of Monsieur Salem, who with his brother stood at the door to welcome us.

The house was a low stone building with two rooms, the first of which served as an entrance hall, the second as a living-room. Unlike Kabyle rooms in general this was furnished with tables, chairs and Oriental rugs, for Salem was a cosmopolitan Kabyle, went often to Algiers, had even been to Chicago, and, although he allowed his wife and children to live in the manner of the country, permitted himself a certain amount of civilsation, not to say luxury, in his personal surroundings. His first question was, when madame would like déjeuner. If madame would name the hour it should be ready, and meantime they would show us the town.

Sight-seeing in a Kabyle village is not arduous work, and after a house or two and the Mosque and Djemaat have been inspected, nothing remains but the people and the view. To get this last they took us to the highest point, whither a considerable number of men and children accompanied us, possibly for the purpose of affording us their protection, more probably to satisfy their curiosity, for foreigners did not visit Ait l'Hassen every day.

If Fort National is called the 'Thorn in the eye of the Kabylie,' the crête upon which the Beni Yenni villages stand might be called the flower in the heart of the Kabylie. It seems to be a centre surrounded by innumerable spurs of about its own height, while the semi-circular wall of Djurjura, towering at no great distance, adds a note of grandeur to which its sublime peaks are forever tuned. In the transparent atmosphere of the Algerian noon but a few steps appeared to separate us from the glittering summit of the Lalla Kredidja, queen of the range.

In walking about the village, the absence of women on the streets was noticeable, while children, many of them handsome, were well represented. Sleek, round-faced, shaven-headed little boys, and merry-eyed, curly-headed girls led the way, ever turning around to gaze shyly at us. The older the girls the blacker their hair, which, from infancy on, is dyed regularly, that it may acquire the degree of jettiness desired by the husband when a girl reaches the marriageable age. Many have blonde hair when born.

Arrived again at Salem's house he shut out all curious visitors and we were soon seated at table. The guides sat cross-legged on the floor at a respectful distance. Two smoking dishes of couscous were brought by a boy, followed by another with a plate of boiled rib of mutton cut in slices. The couscous we were expected to eat with wooden spoons, the round bowls of which were much too large to enter our mouths, and we soon found a certain amount of skill was required to convey the contents of the spoons to the proper destination. The mutton was eaten with the fingers. We were frequently urged to take more by Salem and his brother, who, doubtless in conformity with the customs of hospitality, took nothing themselves, but remained standing ready to serve us, and from time to time deluged our immense dishes of slowly-diminishing couscous with bouillon.

The mixture was not unpalatable, and in smaller portions would have been relished after the long ride; but trying to dispose of quantities large enough for all present was about as hard work as travelling on the mules. Notwithstanding our desire to do honour to the feast, our ardour began to flag.

'You do not eat,' remarked Salem; 'you do not love the couscous?'

'Yes, certainly, we love it above all else, but your portions are very large. We are not accustomed to eat so much.'

'Then take a little more bouillon. That is light and nourishing.'

After one more feeble attempt on the bouillon we gave it up and tried to distract the attention of our friends from ourselves by turning the conversation on their own affairs. This was not difficult, as Salem liked to talk of his experience at the World's Fair at Chicago, whither he had gone in charge of twenty Kabyles under appointment of the Government at Algiers. He seemed particularly impressed, as many a Christian also was, with the Ferris Wheel and the exhibitions in the Midway Plaisance. At length, as if in proof of the truth of his statements, he took out of the inner pockets of his voluminous outer garment a match-box bearing the stars and stripes in red, white and blue, and several other trinkets sold at the Exhibition.

'Ah,' he said with enthusiasm, 'Chicago est une ville magnifique, finer than Beni Yenni, yes, even finer than Algiers.' Did we come from Chicago? No, we did not, and were not sure we preferred it to Beni Yenni. Then followed compliments on American cookery. What could be equal to an American steak? On that point we agreed with him, and heartily wished we had one in place of the farinaceous food before us. But the American coffee was bad, did not compare with café maure. We saw that Salem had good judgment, was an astute observer, and would do credit to the land that had adopted him.

The boy appearing just then with the café maure, our still plenteous rations of mutton and couscous were handed to the patient guides who, with evident appreciation of the quality of the viands, made amends for any want of capacity on our part. Salem and his brother drank coffee with us. While sipping this excellent beverage, we asked Salem where his wife was, and the other women of the family.

'Have patience,' he replied. 'After your coffee you shall see the women.' When we had finished he led the way through a narrow lane to a rear court, where eight or ten women and girls were standing about. With the dignity of a Roman senator he pointed towards them, saying with indifference, 'Voilà les femmes.'

It being the feast of the Ramadan they, like all the women seen the last day or two, were adorned with ornaments. In this case many were of silver, and of rich design. Several of the women were tattooed on the forehead, cheeks and chin, and the hands of all were dyed a deep red. They stood huddled together like a group of shy cattle, children of different ages clinging to their short tunic skirts. The little ones were similarly dyed and decorated. We tried to converse with them, but they only shook their heads and simpered. 'The women understand nothing,' remarked Salem with scorn.

'Which is your wife?' we asked. He looked at the group a moment, then, pointing to a very witch of hideousness, replied, 'Voilà ma femme.'

We went with the women into a typical interior, and examined among other things the earthen dishes used in preparing couscous. Flour forms the basis of this national food eaten by the Kabyles and Arabs usually twice a day. The women roll the flour in their hands until it forms small agglomerations about the size of pearl barley. It is then placed in a low dish, the bottom of which is perforated with little holes, and steamed, no water being allowed to come in contact with it. The poor eat it simply with bouillon, the rich with bouillon and

76

meat. Perhaps, had we realised at breakfast that the henna-dyed hands of Salem's wife had prepared that placed before us, we should have had still less appetite for it.

We left the women and children in their desolate house with its floor of cold earth and returned to Salem's pleasant room with its carpeting of rugs, hoping in our hearts that if our Kabyle friend ever again should go to America, he would study the woman question as well as the culinary excellences of the United States.

Salem showed us a variety of unfinished ornaments, and explained the process of manufacture. As the people were all celebrating the fête we did not see them at work.

The day was passing, so we mounted our mules again and said good-bye to our kind host and Beni Yenni. His brother accompanied us on foot to Fort National. He said he must go on business, and it would give him pleasure to go with us; but we liked to imagine that he was prompted by the chivalrous sentiment of the anaia to see that we reached our destination in safety.

CONCLUSION

AT Fort National our Algerian trip practically ended, for the journey back to Algiers was over ground previously travelled, which offered nothing new. It had been a trip replete with excitement and new experiences. The frequent rains had interfered somewhat with the execution of what had been planned, but enough was accomplished to repay many times all efforts and sacrifices of personal comfort, and to show that the country was well worth visiting, both on account of its natural beauty and the originality of its inhabitants, and also of its wealth in remains of ancient civilisation.

When one has become blasé with years of European travel, let him turn to Algeria and he will find there what will give him new emotions, fresh impressions, enlarge the horizon of his conceptions, and supplement the experiences that have been elsewhere acquired.

THE END

RISE OF DOUAI

RISE OF DOUAI

TWITTER : @RISEOFDOUAI

- AN IDIOT'S GUIDE TO: BEE HUNTING BY JOHN LOCKARD, EDITED BY ARSALAN AHMED (2012)(PAPERBACK) ISBN-10: 1478383550

- THE RISE OF DOUAI BY ARSALAN AHMED (2012)(PAPERBACK) ISBN-10: 1479267783

- THINK AND GROW RICH BY NAPOLEON HILL (2012) (PAPERBACK) ISBN-10: 1480061638
- AS A MAN THINKETH BY MR JAMES ALLEN (2012) (PAPERBACK) ISBN-10: 1480088161
- HOW TO ANALYZE PEOPLE ON SIGHT BY ELSIE BENEDICT (PAPERBACK) ISBN-10: 1480081272
- THE ART OF WAR BY MR SUN TZU (PAPERBACK) ISBN-10: 1480082007
- MACBETH BY MR WILLIAM SHAKESPEARE (PAPERBACK) ISBN-10: 1480060682
- A CHRISTMAS CAROL BY MR CHARLES DICKENS ISBN-10: 1481194755
- CREATING CAPITAL: MONEY-MAKING AS AN AIM IN BUSINESS BY MR FREDRICK L LIPMAN ISBN-10: 1481158996
- GETTING GOLD: A PRACTICAL TREATISE FOR PROSPECTORS, MINERS, AND STUDENTS BY MR JOSEPH COLIN FRANCIS JOHNSON ISBN-10: 1481090984
- THE INTERPRETATION OF DREAMS BY MR SIGMUND FREUD ISBN-10: 1481134558
- THE COMMUNIST MANIFESTO BY MR KARL MARX AND MR FRIDRICK ENGLES ISBN-10: 1480112445
- THE PRINCE BY MR NICCOLO MACHIAVELLI ISBN-10: 1480119601
- THE WAY TO WEALTH BY MR BENJAMIN FRANKLIN

ISBN-10: 1480099651

- THE ART OF MONEY GETTING - OR GOLDEN RULES FOR MAKING MONEY (SUCCESS PRINCIPLES) BY MR PHINEAS TAYLOR BARNUM ISBN-10: 1480138622

Twitter : **@RiseofDouai**

Printed in Great Britain
by Amazon

58295602R00047